Foundations of Data Analysis

Mike Nguyen

Foundations of Data Analysis

Springer

Mike Nguyen
Data Sciences and Operations
University of Southern California
Los Angeles, USA

ISBN 978-3-032-01857-1 ISBN 978-3-032-01858-8 (eBook)
https://doi.org/10.1007/978-3-032-01858-8

Mathematics Subject Classification: 62-00, 62-01, 62F03, 62P05, 62P25, 62A01, 62P20

This Springer imprint is published by the registered company Springer Nature Switzerland AG
The registered company address is: Gewerbestrasse 11, 6330 Cham, Switzerland

Preface

1 Dedication

To the students who ask *why*,
the colleagues who ask *what if*,
and the businesses still learning to listen to their data.
To my family—
for raising me with books, curiosity, and the freedom to ask difficult questions.
You gave me the language of learning long before I knew what to do with it.
To my wife,
whose unwavering belief in me held this book together when I couldn't.
Your love is the constant in a world full of variables.
And finally, to the readers—wherever you are in your learning journey.
May this book illuminate not just how to model data, but how to model the mindset of inquiry, skepticism, and responsible decision-making.
Because data analysis is more than numbers—it's a language for thinking clearly in uncertain times.
This book is for all of you.
For the ones who persist through uncertainty.
For the ones who dare to think statistically.
And for anyone who's ever tried to fit a linear model to a nonlinear life.

2 Acknowledgments

Writing a book is not a solo sport. I may be the name on the cover, but this project exists because of the ideas, patience, caffeine, and generosity of many others.

To my students—you are the heartbeat of this book. Your challenging questions, puzzled expressions, and late-night emails forced me to sharpen ideas until they could withstand real scrutiny. This book owes its structure to you, and its clarity to your confusion. I only hope it gives back half as much as you gave me.

To my mentors—thank you for your wisdom, your patience, and your uncanny ability to ask the one question that dismantled an entire section of my argument (for the better). You showed me how to teach, not just what to teach. Your lectures still echo in my own classroom, and your influence is woven into every chapter of this book.

To the editorial team—your thoughtful feedback, deadlines that flexed just enough, and willingness to let me sneak humor into a statistics book made this process not just manageable, but meaningful. You somehow believed in this book when it was still just a mess of bullet points and half-formed LaTeX tables. You are miracle workers with track changes.

To the open-source R community—my sincerest appreciation. Without `ggplot2`, `dplyr`, and the almighty `bookdown`, this book would still be a loose pile of notebooks and overfit spreadsheets. Your packages did more for reproducible education than I could possibly convey here. If you contributed a pull request to one of the tools I used: this is your book, too.

To my family—you gave me the foundation. You taught me to think critically, read widely, and trust the slow work of learning. You never asked me to explain why I needed another statistics book—you just handed me a bookshelf. Your faith in my intellectual life made this career not only possible but joyful.

And to my wife—you are the quiet center behind this entire project. You carried more than your share of our daily life so I could carry this manuscript across the finish line. You endured my monologues about model assumptions and my rants about inconsistent data structures. You gave me space to think, time to write, and reasons to rest. You reminded me that sometimes the most meaningful models aren't built in R, but in real life—out of patience, laughter, and love.

This book was written in R, but it runs on you.

To everyone listed here, and many more unnamed:

thank you. May this book reflect not only your influence, but your grace.

3 Why This Book?

This book is an effort to simplify and demystify the complex process of data analysis, making it accessible to a wide audience. While I do not claim to be a professional statistician, econometrician, or data scientist, I firmly believe in the value of learning through teaching and practical application. Writing this book has been as much a learning journey for me as I hope it will be for you.

The intended audience includes those with little to no experience in statistics, econometrics, or data science, as well as individuals with a budding interest in these fields who are eager to deepen their knowledge. While my primary domain of interest is marketing, the principles and methods discussed in this book are universally applicable to any discipline that employs scientific methods or data analysis.

I hope this book provides a valuable starting point for aspiring statisticians, econometricians, and data scientists, empowering you to navigate the fascinating world of causal inference and data analysis with confidence.

Los Angeles, CA, USA Mike Nguyen

Contents

Chapter 1
Introduction

Since the turn of the century, we have witnessed remarkable advancements and innovations, particularly in statistics, information technology, computer science, and the rapidly emerging field of data science. However, one challenge of these developments is the overuse of **buzzwords** like *big data*, *machine learning*, and *deep learning*. While these terms are powerful in context, they can sometimes obscure the foundational principles underlying their application.

Every substantive field often has its own specialized metric subfield, such as:

- **Econometrics** in economics
- **Psychometrics** in psychology
- **Chemometrics** in chemistry
- **Sabermetrics** in sports analytics
- **Biostatistics** in public health and medicine

To the layperson, these disciplines are often grouped under broader terms like:

- **Data Science**
- **Applied Statistics**
- **Computational Social Science**

As exciting as it is to explore these new tools and techniques, I must admit that retaining these concepts can be challenging. For me, the most effective way to internalize and apply these ideas has been to document the data analysis process from start to finish.

With that in mind, let's dive in and explore the fascinating world of data analysis together.

M. Nguyen, *Foundations of Data Analysis*,
https://doi.org/10.1007/978-3-032-01858-8_1

1.1 General Recommendations

- The journey of mastering data analysis is fueled by practice and repetition. The more lines of code you write, the more functions you familiarize yourself with, and the more you experiment, the more enjoyable and rewarding this process becomes.
- Readers can approach this book series in several ways:
  - **Focused Learning**: If you are interested in specific methods or tools, you can jump directly to the relevant section by navigating through the table of contents.
  - **Sequential Learning**: To follow a traditional path of data analysis, start with the [Linear Regression] section.
  - **Experimental Approach**: If you are interested in designing experiments and testing hypotheses, explore the [Analysis of Variance (ANOVA)] section.
- For those primarily interested in applications and less concerned with theoretical foundations, focus on the summary and application sections of each chapter.
- If a concept is unclear, consider researching the topic online. This book serves as a guide, and external resources like tutorials or articles can provide additional insights.
- To customize the code examples provided in this book, use R's built-in help functions. For instance:
  - To learn more about a specific function, type `help(function_name)` or `?function_name` in the R console.
  - For example, to find details about the `hist` function, type `?hist` or `help(hist)` in the console.
- Additionally, searching online is a powerful resource (e.g., Google, ChatGPT, etc.). Different practitioners often use various R packages to achieve similar results. For instance, if you need to create a histogram in R, a simple search like *"histogram in R"* will provide multiple approaches and examples.

By adopting these strategies, you can tailor your learning experience and maximize the value of this book.

Tools of Statistics

- Probability Theory
- Mathematical Analysis
- Computer Science
- Numerical Analysis
- Database Management

Code Replication

This book was built with R version 4.4.3 (2025-02-28 ucrt) and the following packages (Table 1.1):

Table 1.1 R packages and versions used

Package	Version	Source
abind	1.4–8	CRAN (R 4.4.1)
agridat	1.24	CRAN (R 4.4.3)
ape	5.8–1	CRAN (R 4.4.3)
assertthat	0.2.1	CRAN (R 4.4.3)
backports	1.5.0	CRAN (R 4.4.0)
bookdown	0.43	CRAN (R 4.4.3)
boot	1.3–31	CRAN (R 4.4.3)
broom	1.0.8	CRAN (R 4.4.3)
bslib	0.9.0	CRAN (R 4.4.2)
cachem	1.1.0	CRAN (R 4.4.2)
callr	3.7.6	CRAN (R 4.4.2)
car	3.1–3	CRAN (R 4.4.2)
carData	3.0–5	CRAN (R 4.4.2)
cellranger	1.1.0	CRAN (R 4.4.2)
cli	3.6.5	CRAN (R 4.4.3)
coda	0.19–4.1	CRAN (R 4.4.2)
colorspace	2.1–1	CRAN (R 4.4.2)
corpcor	1.6.10	CRAN (R 4.4.0)
crayon	1.5.3	CRAN (R 4.4.2)
cubature	2.1.2	CRAN (R 4.4.3)
curl	6.2.2	CRAN (R 4.4.3)
data.table	1.17.2	CRAN (R 4.4.3)
DBI	1.2.3	CRAN (R 4.4.2)
dbplyr	2.5.0	CRAN (R 4.4.2)
desc	1.4.3	CRAN (R 4.4.2)
devtools	2.4.5	CRAN (R 4.4.2)
digest	0.6.37	CRAN (R 4.4.2)
dplyr	1.1.4	CRAN (R 4.4.2)
ellipsis	0.3.2	CRAN (R 4.4.2)
evaluate	1.0.3	CRAN (R 4.4.2)
extrafont	NA	NA
extrafontdb	NA	NA
fansi	1.0.6	CRAN (R 4.4.2)
faraway	1.0.9	CRAN (R 4.4.3)
fastmap	1.2.0	CRAN (R 4.4.2)
forcats	1.0.0	CRAN (R 4.4.2)
foreign	0.8–90	CRAN (R 4.4.3)
fs	1.6.6	CRAN (R 4.4.3)
generics	0.1.4	CRAN (R 4.4.3)
ggplot2	3.5.2	CRAN (R 4.4.3)
glue	1.8.0	CRAN (R 4.4.2)
gtable	0.3.6	CRAN (R 4.4.2)
haven	2.5.4	CRAN (R 4.4.2)

(continued)

Table 1.1 (continued)

Package	Version	Source
Hmisc	5.2–3	CRAN (R 4.4.3)
hms	1.1.3	CRAN (R 4.4.2)
htmltools	0.5.8.1	CRAN (R 4.4.2)
htmlwidgets	1.6.4	CRAN (R 4.4.2)
httr	1.4.7	CRAN (R 4.4.2)
investr	1.4.2	CRAN (R 4.4.3)
jpeg	0.1–11	CRAN (R 4.4.3)
jquerylib	0.1.4	CRAN (R 4.4.2)
jsonlite	2.0.0	CRAN (R 4.4.3)
kableExtra	1.4.0	CRAN (R 4.4.3)
knitr	1.50	CRAN (R 4.4.3)
lattice	0.22–7	CRAN (R 4.4.3)
latticeExtra	0.6–30	CRAN (R 4.4.2)
lifecycle	1.0.4	CRAN (R 4.4.2)
lme4	1.1–37	CRAN (R 4.4.3)
lmerTest	3.1–3	CRAN (R 4.4.2)
lsr	0.5.2	CRAN (R 4.4.3)
ltm	1.2–0	CRAN (R 4.4.3)
lubridate	1.9.4	CRAN (R 4.4.2)
magrittr	2.0.3	CRAN (R 4.4.2)
MASS	7.3–65	CRAN (R 4.4.3)
matlib	1.0.0	CRAN (R 4.4.3)
Matrix	1.7–3	CRAN (R 4.4.3)
MCMCglmm	2.36	CRAN (R 4.4.2)
memoise	2.0.1	CRAN (R 4.4.2)
mgcv	1.9–3	CRAN (R 4.4.3)
minqa	1.2.8	CRAN (R 4.4.2)
modelr	0.1.11	CRAN (R 4.4.2)
munsell	0.5.1	CRAN (R 4.4.2)
nlme	3.1–168	CRAN (R 4.4.3)
nloptr	2.2.1	CRAN (R 4.4.3)
nlstools	2.1–0	CRAN (R 4.4.2)
nnet	7.3–20	CRAN (R 4.4.3)
numDeriv	2016.8–1.1	CRAN (R 4.4.0)
openxlsx	4.2.8	CRAN (R 4.4.3)
pbkrtest	0.5.4	CRAN (R 4.4.3)
pillar	1.10.2	CRAN (R 4.4.3)
pkgbuild	1.4.7	CRAN (R 4.4.3)
pkgconfig	2.0.3	CRAN (R 4.4.2)
pkgload	1.4.0	CRAN (R 4.4.2)
png	0.1–8	CRAN (R 4.4.0)
ppsr	0.0.5	CRAN (R 4.4.3)
prettyunits	1.2.0	CRAN (R 4.4.2)

(continued)

Table 1.1 (continued)

Package	Version	Source
processx	3.8.6	CRAN (R 4.4.3)
ps	1.9.1	CRAN (R 4.4.3)
pscl	1.5.9	CRAN (R 4.4.2)
purrr	1.0.4	CRAN (R 4.4.3)
R6	2.6.1	CRAN (R 4.4.3)
RColorBrewer	1.1–3	CRAN (R 4.4.0)
Rcpp	1.0.14	CRAN (R 4.4.2)
readr	2.1.5	CRAN (R 4.4.2)
readxl	1.4.5	CRAN (R 4.4.3)
remotes	2.5.0	CRAN (R 4.4.2)
reprex	2.1.1	CRAN (R 4.4.2)
rgl	1.3.18	CRAN (R 4.4.3)
rio	1.2.3	CRAN (R 4.4.2)
rlang	1.1.6	CRAN (R 4.4.3)
RLRsim	3.1–8	CRAN (R 4.4.3)
rmarkdown	2.29	CRAN (R 4.4.2)
rprojroot	2.0.4	CRAN (R 4.4.2)
rstudioapi	0.17.1	CRAN (R 4.4.2)
Rttf2pt1	NA	NA
rvest	1.0.4	CRAN (R 4.4.2)
sass	0.4.10	CRAN (R 4.4.3)
scales	1.4.0	CRAN (R 4.4.3)
sessioninfo	1.2.3	CRAN (R 4.4.3)
stringi	1.8.7	CRAN (R 4.4.3)
stringr	1.5.1	CRAN (R 4.4.2)
svglite	2.2.1	CRAN (R 4.4.3)
systemfonts	1.2.3	CRAN (R 4.4.3)
tensorA	0.36.2.1	CRAN (R 4.4.0)
testthat	3.2.3	CRAN (R 4.4.2)
tibble	3.2.1	CRAN (R 4.4.2)
tidyr	1.3.1	CRAN (R 4.4.2)
tidyselect	1.2.1	CRAN (R 4.4.2)
tidyverse	2.0.0	CRAN (R 4.4.2)
tzdb	0.5.0	CRAN (R 4.4.3)
usethis	3.1.0	CRAN (R 4.4.2)
utf8	1.2.5	CRAN (R 4.4.3)
vctrs	0.6.5	CRAN (R 4.4.2)
viridisLite	0.4.2	CRAN (R 4.4.2)
webshot	0.5.5	CRAN (R 4.4.3)
withr	3.0.2	CRAN (R 4.4.2)
xfun	0.52	CRAN (R 4.4.3)
xml2	1.3.8	CRAN (R 4.4.3)
xtable	1.8–4	CRAN (R 4.4.2)
yaml	2.3.10	CRAN (R 4.4.2)
zip	2.3.3	CRAN (R 4.4.3)

```
#> - Session info ------------------------------------------
#>  setting  value
#>  version  R version 4.4.3 (2025-02-28 ucrt)
#>  os       Windows 11 x64 (build 26100)
#>  system   x86_64, mingw32
#>  ui       RTerm
#>  language (EN)
#>  collate  English_United States.utf8
#>  ctype    English_United States.utf8
#>  tz       America/Los_Angeles
#>  date     2025-06-03
#>  pandoc   3.1.1 @ C:/Program Files/RStudio/resources/app/bin/quarto/bin/tools/ (via rmarkdown)
#>  quarto   NA @ C:\\PROGRA~1\\RStudio\\resources\\app\\bin\\quarto\\bin\\quarto.exe
#>
#> - Packages ----------------------------------------------
#>  package     * version date (UTC) lib source
#>  bookdown      0.43    2025-04-15 [1] CRAN (R 4.4.3)
#>  cachem        1.1.0   2024-05-16 [1] CRAN (R 4.4.2)
#>  cli           3.6.5   2025-04-23 [1] CRAN (R 4.4.3)
#>  codetools     0.2-20  2024-03-31 [1] CRAN (R 4.4.3)
#>  desc          1.4.3   2023-12-10 [1] CRAN (R 4.4.2)
#>  devtools      2.4.5   2022-10-11 [1] CRAN (R 4.4.2)
#>  digest        0.6.37  2024-08-19 [1] CRAN (R 4.4.2)
#>  ellipsis      0.3.2   2021-04-29 [1] CRAN (R 4.4.2)
#>  evaluate      1.0.3   2025-01-10 [1] CRAN (R 4.4.2)
#>  fastmap       1.2.0   2024-05-15 [1] CRAN (R 4.4.2)
#>  fs            1.6.6   2025-04-12 [1] CRAN (R 4.4.3)
#>  glue          1.8.0   2024-09-30 [1] CRAN (R 4.4.2)
#>  htmltools     0.5.8.1 2024-04-04 [1] CRAN (R 4.4.2)
#>  htmlwidgets   1.6.4   2023-12-06 [1] CRAN (R 4.4.2)
#>  httpuv        1.6.16  2025-04-16 [1] CRAN (R 4.4.3)
#>  jpeg        * 0.1-11  2025-03-21 [1] CRAN (R 4.4.3)
#>  knitr         1.50    2025-03-16 [1] CRAN (R 4.4.3)
#>  later         1.4.2   2025-04-08 [1] CRAN (R 4.4.3)
#>  lifecycle     1.0.4   2023-11-07 [1] CRAN (R 4.4.2)
#>  magrittr      2.0.3   2022-03-30 [1] CRAN (R 4.4.2)
#>  memoise       2.0.1   2021-11-26 [1] CRAN (R 4.4.2)
#>  mime          0.13    2025-03-17 [1] CRAN (R 4.4.3)
#>  miniUI        0.1.2   2025-04-17 [1] CRAN (R 4.4.3)
#>  pillar        1.10.2  2025-04-05 [1] CRAN (R 4.4.3)
#>  pkgbuild      1.4.7   2025-03-24 [1] CRAN (R 4.4.3)
#>  pkgconfig     2.0.3   2019-09-22 [1] CRAN (R 4.4.2)
#>  pkgload       1.4.0   2024-06-28 [1] CRAN (R 4.4.2)
#>  profvis       0.4.0   2024-09-20 [1] CRAN (R 4.4.2)
#>  promises      1.3.2   2024-11-28 [1] CRAN (R 4.4.2)
#>  purrr         1.0.4   2025-02-05 [1] CRAN (R 4.4.3)
#>  R6            2.6.1   2025-02-15 [1] CRAN (R 4.4.3)
#>  Rcpp          1.0.14  2025-01-12 [1] CRAN (R 4.4.2)
#>  remotes       2.5.0   2024-03-17 [1] CRAN (R 4.4.2)
#>  rlang         1.1.6   2025-04-11 [1] CRAN (R 4.4.3)
#>  rmarkdown     2.29    2024-11-04 [1] CRAN (R 4.4.2)
#>  rstudioapi    0.17.1  2024-10-22 [1] CRAN (R 4.4.2)
#>  sessioninfo   1.2.3   2025-02-05 [1] CRAN (R 4.4.3)
#>  shiny         1.10.0  2024-12-14 [1] CRAN (R 4.4.2)
#>  tibble        3.2.1   2023-03-20 [1] CRAN (R 4.4.2)
#>  tinytex       0.57    2025-04-15 [1] CRAN (R 4.4.3)
#>  urlchecker    1.0.1   2021-11-30 [1] CRAN (R 4.4.2)
#>  usethis       3.1.0   2024-11-26 [1] CRAN (R 4.4.2)
#>  vctrs         0.6.5   2023-12-01 [1] CRAN (R 4.4.2)
#>  xfun          0.52    2025-04-02 [1] CRAN (R 4.4.3)
```

```
#>  xtable        1.8-4   2019-04-21 [1] CRAN (R 4.4.2)
#>  yaml          2.3.10  2024-07-26 [1] CRAN (R 4.4.2)
#>
#>  [1] C:/Program Files/R/R-4.4.3/library
#>  * -- Packages attached to the search path.
#>
#> ----------------------------------------------------------
```

Chapter 2
Prerequisites

This chapter reviews essential mathematical and computational tools required for effective statistical analysis. Beginning with Matrix Theory and core concepts in linear algebra, we then explore foundational Probability Theory, including random variables, distributions, and expectations. The chapter also revisits important mathematical skills—such as functions, derivatives, and logarithms—before moving into practical elements like data import/export and data manipulation using R. These topics ensure that readers are well-equipped to tackle the more advanced statistical methods presented in later chapters.

If you are confident in your understanding of these topics, you can proceed directly to the Descriptive Statistics section to begin exploring applied data analysis.

2.1 Matrix Theory

Matrix A represents the original matrix. It is a 2×2 matrix with elements a_{ij}, where i represents the row and j represents the column.

$$A = \begin{bmatrix} a_{11} & a_{12} \\ a_{21} & a_{22} \end{bmatrix}$$

A' is the transpose of A. The transpose of a matrix flips its rows and columns.

$$A' = \begin{bmatrix} a_{11} & a_{21} \\ a_{12} & a_{22} \end{bmatrix}$$

M. Nguyen, *Foundations of Data Analysis*,
https://doi.org/10.1007/978-3-032-01858-8_2

Fundamental properties and rules of matrices, essential for understanding operations in linear algebra:

$$
\begin{aligned}
(\mathbf{ABC})' &= \mathbf{C}'\mathbf{B}'\mathbf{A}' && \text{(Transpose reverses order in a product)}\\
\mathbf{A}(\mathbf{B}+\mathbf{C}) &= \mathbf{AB}+\mathbf{AC} && \text{(Distributive property)}\\
\mathbf{AB} &\neq \mathbf{BA} && \text{(Multiplication is not commutative)}\\
(\mathbf{A}')' &= \mathbf{A} && \text{(Double transpose is the original matrix)}\\
(\mathbf{A}+\mathbf{B})' &= \mathbf{A}'+\mathbf{B}' && \text{(Transpose of a sum is the sum of transposes)}\\
(\mathbf{AB})' &= \mathbf{B}'\mathbf{A}' && \text{(Transpose reverses order in a product)}\\
(\mathbf{AB})^{-1} &= \mathbf{B}^{-1}\mathbf{A}^{-1} && \text{(Inverse reverses order in a product)}\\
\mathbf{A}+\mathbf{B} &= \mathbf{B}+\mathbf{A} && \text{(Addition is commutative)}\\
\mathbf{AA}^{-1} &= \mathbf{I} && \text{(Matrix times its inverse is identity)}
\end{aligned}
$$

These properties are critical in solving systems of equations, optimizing models, and performing data transformations.

If a matrix **A** has an inverse, it is called *invertible*. If **A** does not have an inverse, it is referred to as *singular*.

The product of two matrices **A** and **B** is computed as

$$
\begin{aligned}
\mathbf{A} &= \begin{bmatrix} a_{11} & a_{12} & a_{13} \\ a_{21} & a_{22} & a_{23} \end{bmatrix} \begin{bmatrix} b_{11} & b_{12} & b_{13} \\ b_{21} & b_{22} & b_{23} \\ b_{31} & b_{32} & b_{33} \end{bmatrix}\\
&= \begin{bmatrix} a_{11}b_{11}+a_{12}b_{21}+a_{13}b_{31} & \sum_{i=1}^{3} a_{1i}b_{i2} & \sum_{i=1}^{3} a_{1i}b_{i3} \\ \sum_{i=1}^{3} a_{2i}b_{i1} & \sum_{i=1}^{3} a_{2i}b_{i2} & \sum_{i=1}^{3} a_{2i}b_{i3} \end{bmatrix}
\end{aligned}
$$

Quadratic Form

Let **a** be a 3×1 vector. The quadratic form involving a matrix **B** is given by

$$\mathbf{a}'\mathbf{B}\mathbf{a} = \sum_{i=1}^{3}\sum_{j=1}^{3} a_i b_{ij} a_j$$

Length of a Vector

The *length* (or 2-norm) of a vector **a**, denoted as $||\mathbf{a}||$, is defined as the square root of the inner product of the vector with itself:

$$||\mathbf{a}|| = \sqrt{\mathbf{a}'\mathbf{a}}$$

2.1.1 Rank of a Matrix

The *rank* of a matrix refers to:

- The dimension of the space spanned by its columns (or rows)
- The number of linearly independent columns or rows

For an $n \times k$ matrix $\mathbf{A}$ and a $k \times k$ matrix $\mathbf{B}$, the following properties hold:

- $\text{rank}(\mathbf{A}) \leq \min(n, k)$.
- $\text{rank}(\mathbf{A}) = \text{rank}(\mathbf{A}') = \text{rank}(\mathbf{A}'\mathbf{A}) = \text{rank}(\mathbf{A}\mathbf{A}')$.
- $\text{rank}(\mathbf{AB}) = \min(\text{rank}(\mathbf{A}), \text{rank}(\mathbf{B}))$.
- $\mathbf{B}$ is invertible (non-singular) if and only if $\text{rank}(\mathbf{B}) = k$.

2.1.2 Inverse of a Matrix

In scalar algebra, if $a = 0$, then $1/a$ does not exist.

In matrix algebra, a matrix is invertible if it is *non-singular*, meaning it has a nonzero determinant and its inverse exists. A square matrix $\mathbf{A}$ is invertible if there exists another square matrix $\mathbf{B}$ such that:

$$\mathbf{AB} = \mathbf{I} \quad \text{(Identity Matrix)}.$$

In this case, $\mathbf{A}^{-1} = \mathbf{B}$.

For a 2×2 matrix:

$$\mathbf{A} = \begin{bmatrix} a & b \\ c & d \end{bmatrix}$$

The inverse is

$$\mathbf{A}^{-1} = \frac{1}{ad - bc} \begin{bmatrix} d & -b \\ -c & a \end{bmatrix}$$

This inverse exists only if $ad - bc \neq 0$, where $ad - bc$ is the determinant of $\mathbf{A}$.

For a partitioned block matrix:

$$\begin{bmatrix} A & B \\ C & D \end{bmatrix}^{-1} = \begin{bmatrix} (\mathbf{A} - \mathbf{BD}^{-1}\mathbf{C})^{-1} & -(\mathbf{A} - \mathbf{BD}^{-1}\mathbf{C})^{-1}\mathbf{BD}^{-1} \\ -\mathbf{D}^{-1}\mathbf{C}(\mathbf{A} - \mathbf{BD}^{-1}\mathbf{C})^{-1} & \mathbf{D}^{-1} + \mathbf{D}^{-1}\mathbf{C}(\mathbf{A} - \mathbf{BD}^{-1}\mathbf{C})^{-1}\mathbf{BD}^{-1} \end{bmatrix}$$

This formula assumes that $\mathbf{D}$ and $\mathbf{A} - \mathbf{BD}^{-1}\mathbf{C}$ are invertible.

Properties of the inverse for non-singular matrices:

1. $(\mathbf{A}^{-1})^{-1} = \mathbf{A}$.
2. For a nonzero scalar b, $(\mathbf{bA})^{-1} = \mathbf{b}^{-1}\mathbf{A}^{-1}$.
3. For a matrix $\mathbf{B}$, $(\mathbf{BA})^{-1} = \mathbf{B}^{-1}\mathbf{A}^{-1}$ (only if $\mathbf{B}$ is non-singular).
4. $(\mathbf{A}^{-1})' = (\mathbf{A}')^{-1}$ (the transpose of the inverse equals the inverse of the transpose).
5. Never notate $\mathbf{1}/\mathbf{A}$; use $\mathbf{A}^{-1}$ instead.

Notes

- The determinant of a matrix determines whether it is invertible. For square matrices, a determinant of 0 means the matrix is *singular* and has no inverse.
- Always verify the conditions for invertibility, particularly when dealing with partitioned or block matrices.

2.1.3 *Definiteness of a Matrix*

A symmetric square $k \times k$ matrix $\mathbf{A}$ is classified based on the following conditions:

- **Positive Semi-definite (PSD)**: $\mathbf{A}$ is PSD if, for any nonzero $k \times 1$ vector $\mathbf{x}$:

$$\mathbf{x}'\mathbf{A}\mathbf{x} \geq \mathbf{0}.$$

- **Negative Semi-definite (NSD)**: $\mathbf{A}$ is NSD if, for any nonzero $k \times 1$ vector $\mathbf{x}$:

$$\mathbf{x}'\mathbf{A}\mathbf{x} \leq \mathbf{0}.$$

- **Indefinite**: $\mathbf{A}$ is indefinite if it is neither PSD nor NSD.

  The *identity matrix* is always positive definite (PD).

Example Let $\mathbf{x} = (x_1, x_2)'$, and consider a 2×2 identity matrix $\mathbf{I}$:

$$\begin{aligned}\mathbf{x}'\mathbf{I}\mathbf{x} &= (x_1, x_2)\begin{bmatrix}1 & 0\\ 0 & 1\end{bmatrix}\begin{bmatrix}x_1\\ x_2\end{bmatrix}\\ &= (x_1, x_2)\begin{bmatrix}x_1\\ x_2\end{bmatrix}\\ &= x_1^2 + x_2^2 \geq 0.\end{aligned}$$

Thus, $\mathbf{I}$ is PD because $\mathbf{x}'\mathbf{I}\mathbf{x} > 0$ for all nonzero $\mathbf{x}$.

Properties of Definiteness

1. Any variance-covariance matrix is PSD.

2. A matrix **A** is PSD if and only if there exists a matrix **B** such that:

$$\mathbf{A} = \mathbf{B}'\mathbf{B}.$$

3. If **A** is PSD, then $\mathbf{B}'\mathbf{A}\mathbf{B}$ is also PSD for any conformable matrix **B**.
4. If **A** and **C** are non-singular, then $\mathbf{A} - \mathbf{C}$ is PSD if and only if $\mathbf{C}^{-1} - \mathbf{A}^{-1}$ is PSD.
5. If **A** is PD (or ND), then $\mathbf{A}^{-1}$ is also PD (or ND).

Notes

- An *indefinite* matrix **A** is neither PSD nor NSD. This concept does not have a direct counterpart in scalar algebra.
- If a square matrix is PSD and invertible, then it is PD.

Examples of Definiteness:

1. **Invertible/Indefinite**:

$$\begin{bmatrix} -1 & 0 \\ 0 & 10 \end{bmatrix}$$

2. **Non-invertible/Indefinite**:

$$\begin{bmatrix} 0 & 1 \\ 0 & 0 \end{bmatrix}$$

3. **Invertible/PSD**:

$$\begin{bmatrix} 1 & 0 \\ 0 & 1 \end{bmatrix}$$

4. **Noninvertible/PSD**:

$$\begin{bmatrix} 0 & 0 \\ 0 & 1 \end{bmatrix}$$

2.1.4 Matrix Calculus

Consider a scalar function $y = f(x_1, x_2, \ldots, x_k) = f(x)$, where x is a $1 \times k$ row vector.

2.1.4.1 Gradient (First-Order Derivative)

The gradient, or the first-order derivative of $f(x)$ with respect to the vector x, is given by

$$\frac{\partial f(x)}{\partial x} = \begin{bmatrix} \frac{\partial f(x)}{\partial x_1} \\ \frac{\partial f(x)}{\partial x_2} \\ \vdots \\ \frac{\partial f(x)}{\partial x_k} \end{bmatrix}$$

2.1.4.2 Hessian (Second-Order Derivative)

The Hessian, or the second-order derivative of $f(x)$ with respect to x, is a symmetric matrix defined as

$$\frac{\partial^2 f(x)}{\partial x \partial x'} = \begin{bmatrix} \frac{\partial^2 f(x)}{\partial x_1^2} & \frac{\partial^2 f(x)}{\partial x_1 \partial x_2} & \cdots & \frac{\partial^2 f(x)}{\partial x_1 \partial x_k} \\ \frac{\partial^2 f(x)}{\partial x_2 \partial x_1} & \frac{\partial^2 f(x)}{\partial x_2^2} & \cdots & \frac{\partial^2 f(x)}{\partial x_2 \partial x_k} \\ \vdots & \vdots & \ddots & \vdots \\ \frac{\partial^2 f(x)}{\partial x_k \partial x_1} & \frac{\partial^2 f(x)}{\partial x_k \partial x_2} & \cdots & \frac{\partial^2 f(x)}{\partial x_k^2} \end{bmatrix}$$

2.1.4.3 Derivative of a Scalar Function with Respect to a Matrix

Let $f(\mathbf{X})$ be a scalar function, where $\mathbf{X}$ is an $n \times p$ matrix. The derivative is

$$\frac{\partial f(\mathbf{X})}{\partial \mathbf{X}} = \begin{bmatrix} \frac{\partial f(\mathbf{X})}{\partial x_{11}} & \cdots & \frac{\partial f(\mathbf{X})}{\partial x_{1p}} \\ \vdots & \ddots & \vdots \\ \frac{\partial f(\mathbf{X})}{\partial x_{n1}} & \cdots & \frac{\partial f(\mathbf{X})}{\partial x_{np}} \end{bmatrix}$$

2.1.4.4 Common Matrix Derivatives

1. If $\mathbf{a}$ is a vector and $\mathbf{A}$ is a matrix independent of $\mathbf{y}$:
   - $\frac{\partial \mathbf{a}'\mathbf{y}}{\partial \mathbf{y}} = \mathbf{a}$.
   - $\frac{\partial \mathbf{y}'\mathbf{y}}{\partial \mathbf{y}} = 2\mathbf{y}$.
   - $\frac{\partial \mathbf{y}'\mathbf{A}\mathbf{y}}{\partial \mathbf{y}} = (\mathbf{A} + \mathbf{A}')\mathbf{y}$.

2. If $\mathbf{X}$ is symmetric:

   - $\frac{\partial|\mathbf{X}|}{\partial x_{ij}} = \begin{cases} X_{ii}, & i = j \\ X_{ij}, & i \neq j, \end{cases}$ where X_{ij} is the (i, j)th cofactor of $\mathbf{X}$.

3. If $\mathbf{X}$ is symmetric and $\mathbf{A}$ is a matrix independent of $\mathbf{X}$:

   - $\frac{\partial \text{tr}(\mathbf{XA})}{\partial \mathbf{X}} = \mathbf{A} + \mathbf{A}' - \text{diag}(\mathbf{A})$.

4. If $\mathbf{X}$ is symmetric, let $\mathbf{J}_{ij}$ be a matrix with 1 at the (i, j)th position and 0 elsewhere:

   - $\frac{\partial \mathbf{X}^{-1}}{\partial x_{ij}} = \begin{cases} -\mathbf{X}^{-1}\mathbf{J}_{ii}\mathbf{X}^{-1}, & i = j \\ -\mathbf{X}^{-1}(\mathbf{J}_{ij} + \mathbf{J}_{ji})\mathbf{X}^{-1}, & i \neq j \end{cases}.$

2.1.5 *Optimization in Scalar and Vector Spaces*

Optimization is the process of finding the minimum or maximum of a function. The conditions for optimization differ depending on whether the function involves a scalar or a vector. Below is a comparison of scalar and vector optimization (Table 2.1):

Key Concepts

1. *First-Order Condition*:

   - The *first-order derivative* of the function must equal zero at a critical point. This holds for both scalar and vector functions:
     - In the scalar case, $\frac{\partial f(x)}{\partial x} = 0$ identifies critical points.
     - In the vector case, $\frac{\partial f(x)}{\partial x}$ is a *gradient vector*, and the condition is satisfied when all elements of the gradient are zero.

Table 2.1 Conditions for scalar and vector optimization

Condition	Scalar optimization	Vector optimization
First-order condition	$\frac{\partial f(x_0)}{\partial x} = 0$	$\frac{\partial f(x_0)}{\partial x} = \begin{bmatrix} 0 \\ \vdots \\ 0 \end{bmatrix}$
Second-order condition For *convex* functions, this implies a *minimum*	$\frac{\partial^2 f(x_0)}{\partial x^2} > 0$	$\frac{\partial^2 f(x_0)}{\partial x \partial x'} > 0$
For *concave* functions, this implies a *maximum*	$\frac{\partial^2 f(x_0)}{\partial x^2} < 0$	$\frac{\partial^2 f(x_0)}{\partial x \partial x'} < 0$

2. *Second-Order Condition*:

   - The *second-order derivative* determines whether the critical point is a minimum, maximum, or saddle point:

     – For scalar functions, $\frac{\partial^2 f(x)}{\partial x^2} > 0$ implies a *local minimum*, while $\frac{\partial^2 f(x)}{\partial x^2} < 0$ implies a *local maximum*.

     – For vector functions, the Hessian matrix $\frac{\partial^2 f(x)}{\partial x \partial x'}$ must be:

       * **Positive Definite**: For a *minimum* (convex function)
       * **Negative Definite**: For a *maximum* (concave function)
       * **Indefinite**: For a *saddle point* (neither minimum nor maximum)

3. **Convex and Concave Functions**:

   - A function $f(x)$ is:

     – *Convex* if $\frac{\partial^2 f(x)}{\partial x^2} > 0$ or the Hessian $\frac{\partial^2 f(x)}{\partial x \partial x'}$ is positive definite.

     – *Concave* if $\frac{\partial^2 f(x)}{\partial x^2} < 0$ or the Hessian is negative definite.

   - Convexity ensures global optimization for minimization problems, while concavity ensures global optimization for maximization problems.

4. **Hessian Matrix**:

   - In vector optimization, the Hessian $\frac{\partial^2 f(x)}{\partial x \partial x'}$ plays a crucial role in determining the nature of critical points:

     – Positive-definite Hessian: All eigenvalues are positive.

     – Negative-definite Hessian: All eigenvalues are negative.

     – Indefinite Hessian: Eigenvalues have mixed signs.

2.1.6 Cholesky Decomposition

In statistical analysis and numerical linear algebra, decomposing matrices into more tractable forms is crucial for efficient computation. One such important factorization is the *Cholesky decomposition*. It applies to *Hermitian* (in the complex case) or *symmetric* (in the real case), *positive-definite* matrices.

Given an $n \times n$ positive-definite matrix A, the Cholesky decomposition states:

$$A = LL^*,$$

where:

- L is a lower-triangular matrix with strictly positive diagonal entries.

- L^* denotes the *conjugate transpose* of L (simply the transpose L^T for real matrices).

Cholesky decomposition is both computationally efficient and numerically stable, making it a go-to technique for many applications—particularly in statistics where we deal extensively with covariance matrices, linear systems, and probability distributions.

Before diving into how we compute a Cholesky decomposition, we need to clarify what it means for a matrix to be *positive definite*. For a real symmetric matrix A:

- A is *positive definite* if for every nonzero vector x, we have

$$x^T A x > 0.$$

- Alternatively, you can characterize positive definiteness by noting that all eigenvalues of A are strictly positive.

Many important matrices in statistics—particularly *covariance* or *precision* matrices—are both symmetric and positive definite.

2.1.6.1 Existence

A real $n \times n$ matrix A that is symmetric and positive definite always admits a Cholesky decomposition $A = LL^T$. This theorem guarantees that for any covariance matrix in statistics—assuming it is valid (i.e., positive definite)—we can decompose it via Cholesky.

2.1.6.2 Uniqueness

If we additionally require that the diagonal entries of L are *strictly positive*, then L is *unique*. That is, no other lower-triangular matrix with strictly positive diagonal entries will produce the same factorization. This uniqueness is helpful for ensuring consistent numerical outputs in software implementations.

2.1.6.3 Constructing the Cholesky Factor *L*

Given a real, symmetric, positive-definite matrix $A \in \mathbb{R}^{n \times n}$, we want to find the lower-triangular matrix L such that $A = LL^T$. One way to do this is by using a simple step-by-step procedure (often part of standard linear algebra libraries):

1. *Initialize* L to be an $n \times n$ zero matrix.
2. *Iterate* through the rows $i = 1, 2, \ldots, n$:

1. For each row i, compute

$$L_{ii} = \sqrt{A_{ii} - \sum_{k=1}^{i-1} L_{ik}^2}.$$

2. For each column $j = i + 1, i + 2, \ldots, n$:

$$L_{ji} = \frac{1}{L_{ii}} \left(A_{ji} - \sum_{k=1}^{i-1} L_{jk} L_{ik} \right).$$

3. All other entries of L remain zero or are computed in subsequent steps.

3. **Result**: L is lower triangular, and L^T is its transpose.

Cholesky decomposition is roughly half the computational cost of a more general LU decomposition. Specifically, it requires on the order of $\frac{1}{3}n^3$ floating-point operations (flops), making it significantly more efficient in practice than other decompositions for positive-definite systems.

2.1.6.4 Illustrative Example

Consider a small 3×3 positive-definite matrix:

$$A = \begin{pmatrix} 4 & 2 & 4 \\ 2 & 5 & 6 \\ 4 & 6 & 20 \end{pmatrix}.$$

We claim A is positive definite (one could check by calculating principal minors or verifying $x^T Ax > 0$ for all $x \neq 0$). We find L step by step:

1. *Compute* L_{11}:

$$L_{11} = \sqrt{A_{11}} = \sqrt{4} = 2.$$

2. *Compute* L_{21} and L_{31}:
   - $L_{21} = \frac{A_{21}}{L_{11}} = \frac{2}{2} = 1.$
   - $L_{31} = \frac{A_{31}}{L_{11}} = \frac{4}{2} = 2.$

3. *Compute* L_{22}:

$$L_{22} = \sqrt{A_{22} - L_{21}^2} = \sqrt{5 - 1^2} = \sqrt{4} = 2.$$

4. *Compute* L_{32}:

$$L_{32} = \frac{A_{32} - L_{31}L_{21}}{L_{22}} = \frac{6 - (2)(1)}{2} = \frac{4}{2} = 2.$$

5. *Compute* L_{33}:

$$L_{33} = \sqrt{A_{33} - (L_{31}^2 + L_{32}^2)} = \sqrt{20 - (2^2 + 2^2)} = \sqrt{20 - 8} = \sqrt{12} = 2\sqrt{3}.$$

Thus,

$$L = \begin{pmatrix} 2 & 0 & 0 \\ 1 & 2 & 0 \\ 2 & 2 & 2\sqrt{3} \end{pmatrix}.$$

One can verify $LL^T = A$.

2.1.6.5 Applications in Statistics

2.1.6.5.1 Solving Linear Systems

A common statistical problem is solving $Ax = b$ for x. For instance, in regression or in computing Bayesian posterior modes, we often need to solve linear equations with covariance or precision matrices. With $A = LL^T$:

1. **Forward Substitution**: Solve $L\,y = b$.
2. **Backward Substitution**: Solve $L^T x = y$.

This two-step process is more stable and efficient than directly inverting A (which is typically discouraged due to numerical issues).

2.1.6.5.2 Generating Correlated Random Vectors

In simulation-based statistics (e.g., Monte Carlo methods), we often need to generate random draws from a *multivariate normal* distribution $\mathcal{N}(\mu, \Sigma)$, where Σ is the covariance matrix. The steps are:

1. Generate a vector $z \sim \mathcal{N}(0, I)$ of independent standard normal variables.
2. Compute $x = \mu + Lz$, where $\Sigma = LL^T$.

Then x has the desired covariance structure Σ. This technique is widely used in Bayesian statistics (e.g., MCMC sampling) and financial modeling (e.g., portfolio simulations).

2.1.6.5.3 Gaussian Processes and Kriging

In Gaussian process modeling (common in spatial statistics, machine learning, and geostatistics), we frequently work with large covariance matrices that describe the correlations between observed data points:

$$\Sigma = \begin{pmatrix} k(x_1, x_1) & k(x_1, x_2) & \cdots & k(x_1, x_n) \\ k(x_2, x_1) & k(x_2, x_2) & \cdots & k(x_2, x_n) \\ \vdots & \vdots & \ddots & \vdots \\ k(x_n, x_1) & k(x_n, x_2) & \cdots & k(x_n, x_n) \end{pmatrix},$$

where $k(\cdot, \cdot)$ is a covariance (kernel) function. We may need to invert or factorize Σ repeatedly to evaluate the log likelihood:

$$\log \mathcal{L}(\theta) \sim -\tfrac{1}{2} (y - m(\theta))^T \Sigma^{-1} (y - m(\theta)) - \tfrac{1}{2} \log |\Sigma|,$$

where $m(\theta)$ is the mean function and θ are parameters. Using the Cholesky factor L of Σ helps:

- Σ^{-1} can be implied by solving systems with L instead of explicitly computing the inverse.
- $\log |\Sigma|$ can be computed as $2 \sum_{i=1}^{n} \log L_{ii}$.

Hence, Cholesky decomposition becomes the backbone of Gaussian process computations.

2.1.6.5.4 Bayesian Inference with Covariance Matrices

Many Bayesian models—especially hierarchical models—assume a multivariate normal prior on parameters. Cholesky decomposition is used to:

- Sample from these priors or from posterior distributions.
- Regularize large covariance matrices.
- Speed up Markov Chain Monte Carlo (MCMC) computations by factorizing covariance structures.

2.1.6.6 Other Notes

1. Numerical Stability Considerations

Cholesky decomposition is considered more stable than a general LU decomposition when applied to positive-definite matrices. Since no row or column pivots are required, rounding errors can be smaller. Of course, in practice, software implementations can vary, and extremely ill-conditioned matrices can still pose numerical challenges:

2. Why We *Do Not* Usually Compute $\mathbf{A}^{-1}$?

It is common in statistics (especially in older texts) to see formulas involving Σ^{-1}. However, computing an inverse explicitly is often discouraged because:

- It is numerically less stable.
- It requires more computations.
- Many tasks that *appear* to need Σ^{-1} can be done more efficiently by solving systems via the Cholesky factor L.

Hence, "*solve, don't invert*" is a common mantra. If you see an expression like $\Sigma^{-1}b$, you can use the Cholesky factors L and L^T to solve $\Sigma x = b$ by forward and backward substitution, bypassing the direct inverse calculation:

3. Further Variants and Extensions:

- **Incomplete Cholesky**: Sometimes used in iterative solvers where a full Cholesky factorization is too expensive, especially for large sparse systems.
- **LDL^T Decomposition**: A variant that avoids taking square roots; used for positive *semi*-definite or indefinite systems, but with caution about pivoting strategies.

2.2 Probability Theory

2.2.1 Axioms and Theorems of Probability

1. Let S denote the sample space of an experiment. Then:

$$P[S] = 1.$$

(The probability of the sample space is always 1.)

2. For any event A:

$$P[A] \geq 0.$$

(Probabilities are always nonnegative.)

3. Let $A_1, A_2, A_3, \ldots$ be a finite or infinite collection of mutually exclusive events. Then:

$$P[A_1 \cup A_2 \cup A_3 \ldots] = P[A_1] + P[A_2] + P[A_3] + \ldots$$

(The probability of the union of mutually exclusive events is the sum of their probabilities.)

4. The probability of the empty set is:

$$P[\emptyset] = 0.$$

5. The complement rule:

$$P[A'] = 1 - P[A].$$

6. The probability of the union of two events:

$$P[A_1 \cup A_2] = P[A_1] + P[A_2] - P[A_1 \cap A_2].$$

2.2.1.1 Conditional Probability

The conditional probability of A given B is defined as

$$P[A|B] = \frac{P[A \cap B]}{P[B]}, \quad \text{provided } P[B] \neq 0.$$

2.2.1.2 Independent Events

Two events A and B are independent if and only if:

1. $P[A \cap B] = P[A]P[B]$.
2. $P[A|B] = P[A]$.
3. $P[B|A] = P[B]$.

A collection of events $A_1, A_2, \ldots, A_n$ is independent if and only if every subcollection is independent.

2.2.1.3 Multiplication Rule

The probability of the intersection of two events can be calculated as:

$$P[A \cap B] = P[A|B]P[B] = P[B|A]P[A].$$

2.2.1.4 Bayes' Theorem

Let $A_1, A_2, \ldots, A_n$ be a collection of mutually exclusive events whose union is S, and let B be an event with $P[B] \neq 0$. Then, for any event A_j $(j = 1, 2, \ldots, n)$:

$$P[A_j|B] = \frac{P[B|A_j]P[A_j]}{\sum_{i=1}^{n} P[B|A_i]P[A_i]}.$$

2.2.1.5 Jensen's Inequality

- If $g(x)$ is convex, then:

$$E[g(X)] \geq g(E[X]).$$

- If $g(x)$ is concave, then:

$$E[g(X)] \leq g(E[X]).$$

Jensen's inequality provides a useful way to demonstrate why the standard error calculated using the sample standard deviation (s) as a proxy for the population standard deviation (σ) is a biased estimator:

- The population standard deviation σ is defined as:

$$\sigma = \sqrt{\mathbb{E}[(X - \mu)^2]},$$

  where $\mu = \mathbb{E}[X]$ is the population mean.
- The sample standard deviation s is given by:

$$s = \sqrt{\frac{1}{n-1}\sum_{i=1}^{n}(X_i - \bar{X})^2},$$

  where $\bar{X}$ is the sample mean.
- When s is used as an estimator for σ, the expectation involves the square root function, which is concave.

Applying Jensen's Inequality
The standard error formula involves the square root:

$$\sqrt{\mathbb{E}[s^2]}.$$

However, because the square root function is concave, Jensen's inequality implies:

$$\sqrt{\mathbb{E}[s^2]} \leq \mathbb{E}[\sqrt{s^2}] = \mathbb{E}[s].$$

This inequality shows that the expected value of s (the sample standard deviation) systematically underestimates the population standard deviation σ.

Quantifying the Bias
The bias arises because:

$$\mathbb{E}[s] \neq \sigma.$$

To correct this bias, we note that the sample standard deviation is related to the population standard deviation by:

$$\mathbb{E}[s] = \sigma \cdot \sqrt{\frac{n-1}{n}},$$

where n is the sample size. This bias decreases as n increases, and the estimator becomes asymptotically unbiased.

By leveraging Jensen's inequality, we observe that the concavity of the square root function ensures that s is a biased estimator of σ, systematically underestimating the population standard deviation.

2.2.1.6 Law of Iterated Expectation

The *Law of Iterated Expectation* states that for random variables X and Y:

$$E(X) = E(E(X|Y)).$$

This means the expected value of X can be obtained by first calculating the conditional expectation $E(X|Y)$ and then taking the expectation of this quantity over the distribution of Y.

2.2.1.7 Correlation and Independence

The strength of the relationship between random variables can be ranked from strongest to weakest as:

1. **Independence**:
   - $f(x, y) = f_X(x) f_Y(y)$.

- $f_{Y|X}(y|x) = f_Y(y)$ and $f_{X|Y}(x|y) = f_X(x)$.
- $E[g_1(X)g_2(Y)] = E[g_1(X)]E[g_2(Y)]$.

2. **Mean Independence** (implied by independence):
   - Y is mean independent of X if:

$$E[Y|X] = E[Y].$$

   - $E[Xg(Y)] = E[X]E[g(Y)]$.

3. *Uncorrelatedness* (implied by independence and mean independence):
   - $\text{Cov}(X, Y) = 0$.
   - $\text{Var}(X + Y) = \text{Var}(X) + \text{Var}(Y)$.
   - $E[XY] = E[X]E[Y]$.

2.2.2 *Central Limit Theorem*

The *central limit theorem* states that for a sufficiently large sample size ($n \geq 25$), the sampling distribution of the sample mean or proportion approaches a normal distribution, regardless of the population's original distribution.

Let $X_1, X_2, \ldots, X_n$ be a random sample of size n from a distribution X with mean μ and variance σ^2. Then, for large n:

1. The sample mean $\bar{X}$ is approximately normal:

$$\mu_{\bar{X}} = \mu, \quad \sigma_{\bar{X}}^2 = \frac{\sigma^2}{n}.$$

2. The sample proportion $\hat{p}$ is approximately normal:

$$\mu_{\hat{p}} = p, \quad \sigma_{\hat{p}}^2 = \frac{p(1-p)}{n}.$$

3. The difference in sample proportions $\hat{p}_1 - \hat{p}_2$ is approximately normal:

$$\mu_{\hat{p}_1-\hat{p}_2} = p_1 - p_2, \quad \sigma_{\hat{p}_1-\hat{p}_2}^2 = \frac{p_1(1-p_1)}{n_1} + \frac{p_2(1-p_2)}{n_2}.$$

4. The difference in sample means $\bar{X}_1 - \bar{X}_2$ is approximately normal:

$$\mu_{\bar{X}_1-\bar{X}_2} = \mu_1 - \mu_2, \quad \sigma_{\bar{X}_1-\bar{X}_2}^2 = \frac{\sigma_1^2}{n_1} + \frac{\sigma_2^2}{n_2}.$$

5. The following random variables are approximately standard normal:

- $\frac{\bar{X}-\mu}{\sigma/\sqrt{n}}$.
- $\frac{\hat{p}-p}{\sqrt{\frac{p(1-p)}{n}}}$.
- $\frac{(\hat{p}_1-\hat{p}_2)-(p_1-p_2)}{\sqrt{\frac{p_1(1-p_1)}{n_1}+\frac{p_2(1-p_2)}{n_2}}}$.
- $\frac{(\bar{X}_1-\bar{X}_2)-(\mu_1-\mu_2)}{\sqrt{\frac{\sigma_1^2}{n_1}+\frac{\sigma_2^2}{n_2}}}$.

2.2.2.1 Limiting Distribution of the Sample Mean

If $\{X_i\}_{i=1}^n$ is an iid random sample from a distribution with finite mean μ and finite variance σ^2, the sample mean $\bar{X}$ scaled by $\sqrt{n}$ has the following limiting distribution:

$$\sqrt{n}(\bar{X}-\mu) \xrightarrow{d} N(0, \sigma^2).$$

Standardizing the sample mean gives:

$$\frac{\sqrt{n}(\bar{X}-\mu)}{\sigma} \xrightarrow{d} N(0, 1).$$

Notes

- The CLT holds for most random samples from any distribution (continuous, discrete, or unknown).
- It extends to the multivariate case: A random sample of a random vector converges to a multivariate normal distribution.

2.2.2.2 Asymptotic Variance and Limiting Variance

Asymptotic Variance (Avar):

$$Avar(\sqrt{n}(\bar{X}-\mu)) = \sigma^2.$$

- Refers to the variance of the limiting distribution of an estimator as the sample size (n) approaches infinity.
- It characterizes the variability of the scaled estimator $\sqrt{n}(\bar{x}-\mu)$ in its asymptotic distribution (e.g., normal distribution).

Limiting Variance ($\lim_{n\to\infty} Var$)

$$\lim_{n\to\infty} Var(\sqrt{n}(\bar{x}-\mu)) = \sigma^2$$

- Represents the value that the actual variance of $\sqrt{n}(\bar{x}-\mu)$ converges to as $n \to \infty$.

  For a well-behaved estimator,

$$Avar(\sqrt{n}(\bar{X}-\mu)) = \lim_{n\to\infty} Var(\sqrt{n}(\bar{x}-\mu)) = \sigma^2.$$

However, *asymptotic variance is not necessarily equal to the limiting value of the variance* because asymptotic variance is derived from the limiting distribution, while limiting variance is a convergence result of the sequence of variances.

$$Avar(.) \neq lim_{n\to\infty} Var(.)$$

- Both the asymptotic variance $Avar$ and the limiting variance $\lim_{n\to\infty} Var$ are numerically equal to σ^2, but their conceptual definitions differ.
- $Avar(\cdot) \neq \lim_{n\to\infty} Var(\cdot)$. This emphasizes that while the numerical result may match, their derivation and meaning differ:
  - $Avar$ depends on the asymptotic (large sample) distribution of the estimator.
  - $\lim_{n\to\infty} Var(\cdot)$ involves the sequence of variances as n grows.

  Cases where the two do not match:

1. **Sample Quantiles**: Consider the sample quantile of order p, for some $0 < p < 1$. Under regularity conditions, the asymptotic distribution of the sample quantile is normal, with a variance that depends on p and the density of the distribution at the pth quantile. However, the variance of the sample quantile itself does not necessarily converge to this limit as the sample size grows.
2. **Bootstrap Methods**: When using bootstrapping techniques to estimate the distribution of a statistic, the bootstrap distribution might converge to a different limiting distribution than the original statistic. In these cases, the variance of the bootstrap distribution (or the bootstrap variance) might differ from the limiting variance of the original statistic.
3. **Statistics with Randomly Varying Asymptotic Behavior**: In some cases, the asymptotic behavior of a statistic can vary randomly depending on the sample path. For such statistics, the asymptotic variance might not provide a consistent estimate of the limiting variance.
4. **M-estimators with Varying Asymptotic Behavior**: M-estimators can sometimes have different asymptotic behaviors depending on the tail behavior of

the underlying distribution. For heavy-tailed distributions, the variance of the estimator might not stabilize even as the sample size grows large, making the asymptotic variance different from the variance of any limiting distribution.

2.2.3 *Random Variable*

Random variables can be categorized as either *discrete* or *continuous*, with distinct properties and functions defining each type (Table 2.2).

Expected Value Properties

- $E[c] = c$ for any constant c.
- $E[cX] = cE[X]$ for any constant c.
- $E[X + Y] = E[X] + E[Y]$.
- $E[XY] = E[X]E[Y]$ (if X and Y are independent).

Variance Properties:

- $\mathrm{Var}(c) = 0$ for any constant c.
- $\mathrm{Var}(cX) = c^2\mathrm{Var}(X)$ for any constant c.
- $\mathrm{Var}(X) \geq 0$.
- $\mathrm{Var}(X) = E[X^2] - (E[X])^2$.
- $\mathrm{Var}(X + c) = \mathrm{Var}(X)$.
- $\mathrm{Var}(X + Y) = \mathrm{Var}(X) + \mathrm{Var}(Y)$ (if X and Y are independent).

Table 2.2 Comparison of discrete and continuous random variables

	Discrete variable	Continuous variable
Definition	A random variable is discrete if it can assume at most a finite or countably infinite number of values	A random variable is continuous if it can assume any value in some interval or intervals of real numbers, with $P(X = x) = 0$
Density function	A function f is called a density for X if:	A function f is called a density for X if:
	1. $f(x) \geq 0$	1. $f(x) \geq 0$ for x real
	2. $\sum_x f(x) = 1$	2. $\int_{-\infty}^{\infty} f(x)\,dx = 1$
	3. $f(x) = P(X = x)$ for x real	3. $P[a \leq X \leq b] = \int_a^b f(x)\,dx$ for a, b real
Cumulative distribution function	$F(x) = P(X \leq x)$	$F(x) = P(X \leq x) = \int_{-\infty}^{x} f(t)\,dt$
$E[H(X)]$	$\sum_x H(x) f(x)$	$\int_{-\infty}^{\infty} H(x) f(x)\,dx$
$\mu = E[X]$	$\sum_x x f(x)$	$\int_{-\infty}^{\infty} x f(x)\,dx$
Ordinary moments	$\sum_x x^k f(x)$	$\int_{-\infty}^{\infty} x^k f(x)\,dx$
Moment generating function	$m_X(t) = E[e^{tX}] = \sum_x e^{tx} f(x)$	$m_X(t) = E[e^{tX}] = \int_{-\infty}^{\infty} e^{tx} f(x)\,dx$

The standard deviation σ is given by:

$$\sigma = \sqrt{\sigma^2} = \sqrt{\text{Var}(X)}.$$

2.2.3.1 Multivariate Random Variables

Suppose $y_1, \ldots, y_p$ are random variables with means $\mu_1, \ldots, \mu_p$. Then:

$$\mathbf{y} = \begin{bmatrix} y_1 \\ \vdots \\ y_p \end{bmatrix}, \quad E[\mathbf{y}] = \begin{bmatrix} \mu_1 \\ \vdots \\ \mu_p \end{bmatrix} = \boldsymbol{\mu}.$$

The covariance between y_i and y_j is $\sigma_{ij} = \text{Cov}(y_i, y_j)$. The variance-covariance (or dispersion) matrix is

$$\boldsymbol{\Sigma} = (\sigma_{ij}) = \begin{bmatrix} \sigma_{11} & \sigma_{12} & \ldots & \sigma_{1p} \\ \sigma_{21} & \sigma_{22} & \ldots & \sigma_{2p} \\ \vdots & \vdots & \ddots & \vdots \\ \sigma_{p1} & \sigma_{p2} & \ldots & \sigma_{pp} \end{bmatrix}$$

And $\boldsymbol{\Sigma}$ is symmetric with $(p+1)p/2$ unique parameters.

Alternatively, let $u_{p\times 1}$ and $v_{v\times 1}$ be random vectors with means $\boldsymbol{\mu}_{\mathbf{u}}$ and $\boldsymbol{\mu}_{\mathbf{v}}$. Then

$$\boldsymbol{\Sigma}_{\mathbf{uv}} = cov(\mathbf{u}, \mathbf{v}) = E[(\mathbf{u} - \boldsymbol{\mu}_{\mathbf{u}})(\mathbf{v} - \boldsymbol{\mu}_{\mathbf{v}})']$$

$\Sigma_{uv} \neq \Sigma_{vu}$ (but $\Sigma_{uv} = \Sigma'_{vu}$).

Properties of Covariance Matrices

1. **Symmetry**: $\boldsymbol{\Sigma}' = \boldsymbol{\Sigma}$.
2. **Eigendecomposition** (spectral decomposition, symmetric decomposition): $\boldsymbol{\Sigma} = \boldsymbol{\Phi}\boldsymbol{\Lambda}\boldsymbol{\Phi}$, where $\boldsymbol{\Phi}$ is a matrix of eigenvectors such that $\boldsymbol{\Phi}\boldsymbol{\Phi}' = \mathbf{I}$ (orthonormal), and $\boldsymbol{\Lambda}$ is a diagonal matrix with eigenvalues $(\lambda_1, \ldots, \lambda_p)$ on the diagonal.
3. **Nonnegative Definiteness**: $\mathbf{a}\boldsymbol{\Sigma}\mathbf{a} \geq 0$ for any $\mathbf{a} \in R^p$. Equivalently, the eigenvalues of $\boldsymbol{\Sigma}$, $\lambda_1 \geq \ldots \geq \lambda_p \geq 0$.
4. **Generalized Variance**: $|\boldsymbol{\Sigma}| = \lambda_1 \ldots \lambda_p \geq 0$.
5. **Trace**: $\text{tr}(\boldsymbol{\Sigma}) = \lambda_1 + \cdots + \lambda_p = \sigma_{11} + \cdots + \sigma_{pp} = \sum \sigma_{ii}$ = sum of variances (total variance).

Note $\boldsymbol{\Sigma}$ is required to be positive definite. This implies that all eigenvalues are positive, and $\boldsymbol{\Sigma}$ has an inverse $\boldsymbol{\Sigma}^{-1}$, such that $\boldsymbol{\Sigma}^{-1}\boldsymbol{\Sigma} = \mathbf{I}_{p\times p} = \boldsymbol{\Sigma}\boldsymbol{\Sigma}^{-1}$.

2.2.3.2 Correlation Matrices

The correlation coefficient ρ_{ij} and correlation matrix $\mathbf{R}$ are defined as

$$\rho_{ij} = \frac{\sigma_{ij}}{\sqrt{\sigma_{ii}\sigma_{jj}}}, \qquad \mathbf{R} = \begin{bmatrix} 1 & \rho_{12} & \dots & \rho_{1p} \\ \rho_{21} & 1 & \dots & \rho_{2p} \\ \vdots & \vdots & \ddots & \vdots \\ \rho_{p1} & \rho_{p2} & \dots & 1 \end{bmatrix}$$

where $\rho_{ii} = 1 \forall i$.

2.2.3.3 Linear Transformations

Let $\mathbf{A}$ and $\mathbf{B}$ be matrices of constants, and $\mathbf{c}$ and $\mathbf{d}$ be vectors of constants. Then:

- $E[\mathbf{Ay} + \mathbf{c}] = \mathbf{A}\boldsymbol{\mu}_\mathbf{y} + \mathbf{c}$.
- $\text{Var}(\mathbf{Ay} + \mathbf{c}) = \mathbf{A}\boldsymbol{\Sigma}_\mathbf{y}\mathbf{A}'$.
- $\text{Cov}(\mathbf{Ay} + \mathbf{c}, \mathbf{By} + \mathbf{d}) = \mathbf{A}\boldsymbol{\Sigma}_\mathbf{y}\mathbf{B}'$.

2.2.4 *Moment Generating Function*

2.2.4.1 Properties of the Moment Generating Function

1. $\left.\frac{d^k(m_X(t))}{dt^k}\right|_{t=0} = E[X^k]$. (The kth derivative at $t = 0$ gives the kth moment of X.)
2. $\mu = E[X] = m'_X(0)$. (The first derivative at $t = 0$ gives the mean.)
3. $E[X^2] = m''_X(0)$. (The second derivative at $t = 0$ gives the second moment.)

2.2.4.2 Theorems Involving MGFs

Let $X_1, X_2, \dots, X_n, Y$ be random variables with MGFs $m_{X_1}(t), m_{X_2}(t), \dots, m_{X_n}(t), m_Y(t)$:

1. If $m_{X_1}(t) = m_{X_2}(t)$ for all t in some open interval about 0, then X_1 and X_2 have the same distribution.
2. If $Y = \alpha + \beta X_1$, then:

$$m_Y(t) = e^{\alpha t} m_{X_1}(\beta t).$$

3. If $X_1, X_2, \ldots, X_n$ are independent and $Y = \alpha_0 + \alpha_1 X_1 + \alpha_2 X_2 + \cdots + \alpha_n X_n$, where $\alpha_0, \alpha_1, \ldots, \alpha_n$ are constants, then:

$$m_Y(t) = e^{\alpha_0 t} m_{X_1}(\alpha_1 t) m_{X_2}(\alpha_2 t) \ldots m_{X_n}(\alpha_n t).$$

4. Suppose $X_1, X_2, \ldots, X_n$ are independent normal random variables with means $\mu_1, \mu_2, \ldots, \mu_n$ and variances $\sigma_1^2, \sigma_2^2, \ldots, \sigma_n^2$. If $Y = \alpha_0 + \alpha_1 X_1 + \alpha_2 X_2 + \cdots + \alpha_n X_n$, then:

   - Y is normally distributed.
   - Mean: $\mu_Y = \alpha_0 + \alpha_1 \mu_1 + \alpha_2 \mu_2 + \cdots + \alpha_n \mu_n$.
   - Variance: $\sigma_Y^2 = \alpha_1^2 \sigma_1^2 + \alpha_2^2 \sigma_2^2 + \cdots + \alpha_n^2 \sigma_n^2$.

2.2.5 *Moments*

See Table 2.3

- **Skewness**: Skewness$(X) = \frac{E[(X-\mu)^3]}{\sigma^3}$.
  - **Definition**: Skewness measures the asymmetry of a probability distribution around its mean.
  - **Interpretation**:
    * **Positive skewness**: The right tail (higher values) is longer or heavier than the left tail.
    * **Negative skewness**: The left tail (lower values) is longer or heavier than the right tail.
    * **Zero skewness**: The data is symmetric.

- **Kurtosis**: Kurtosis$(X) = \frac{E[(X-\mu)^4]}{\sigma^4}$.
  - **Definition**: Kurtosis measures the "tailedness" or the heaviness of the tails of a probability distribution.
  - Excess kurtosis (often reported) is the kurtosis minus 3 (to compare against the normal distribution's kurtosis of 3).

Table 2.3 First four moments of a random variable, centered and uncentered

Moment	Uncentered	Centered
1st	$E[X] = \mu = \text{Mean}(X)$	
2nd	$E[X^2]$	$E[(X-\mu)^2] = \text{Var}(X) = \sigma^2$
3rd	$E[X^3]$	$E[(X-\mu)^3]$
4th	$E[X^4]$	$E[(X-\mu)^4]$

- **Interpretation**:
  * **High kurtosis (>3)**: Heavy tails, more extreme outliers
  * **Low kurtosis (<3)**: Light tails, fewer outliers
  * **Normal distribution kurtosis = 3**: Benchmark for comparison

2.2.6 *Skewness*

Skewness measures the asymmetry of the distribution:

1. **Positive skew**: The right side (high values) is stretched out.
   - Positive skew occurs when the right tail (higher values) of the distribution is longer or heavier.
   - **Examples**:
     - **Income Distribution**: In many countries, most people earn a moderate income, but a small fraction of ultra-high earners stretches the distribution's right tail.
     - **Housing Prices**: Most homes may be around an affordable price, but a few extravagant mansions create a very long (and expensive) upper tail (Fig. 2.1).

```r
# Load required libraries
library(ggplot2)

# Simulate data for positive skew
set.seed(123)
positive_skew_income <-
    rbeta(1000, 5, 2) * 100  # Income distribution example
positive_skew_housing <-
    rbeta(1000, 5, 2) * 1000  # Housing prices example

# Combine data
data_positive_skew <- data.frame(
    value = c(positive_skew_income, positive_skew_housing),
    example = c(rep("Income Distribution", 1000),
                rep("Housing Prices", 1000))
)

# Plot positive skew
ggplot(data_positive_skew, aes(x = value, fill = example)) +
    geom_histogram(bins = 30,
                   alpha = 0.7,
                   position = "identity") +
    facet_wrap( ~ example, scales = "free") +
    labs(title = "Visualization of Positive Skew",
```

```
       x = "Value",
       y = "Frequency") +
  theme_minimal()
```

- In the *income distribution* example, most people earn moderate incomes, but a few high earners stretch the right tail.
- In the *Housing Prices* example, most homes are reasonably priced, but a few mansions create a long, expensive right tail.

2. Negative skew (left skew):

- Negative skew occurs when the left tail (lower values) of the distribution is longer or heavier.
- **Examples**:
  - **Scores on an Easy Test**: If an exam is very easy, most students score quite high, and only a few students score low, creating a left tail.
  - **Age of Retirement**: Most people might retire around a common age (say 65+), with fewer retiring very early (stretching the left tail) (Fig. 2.2).

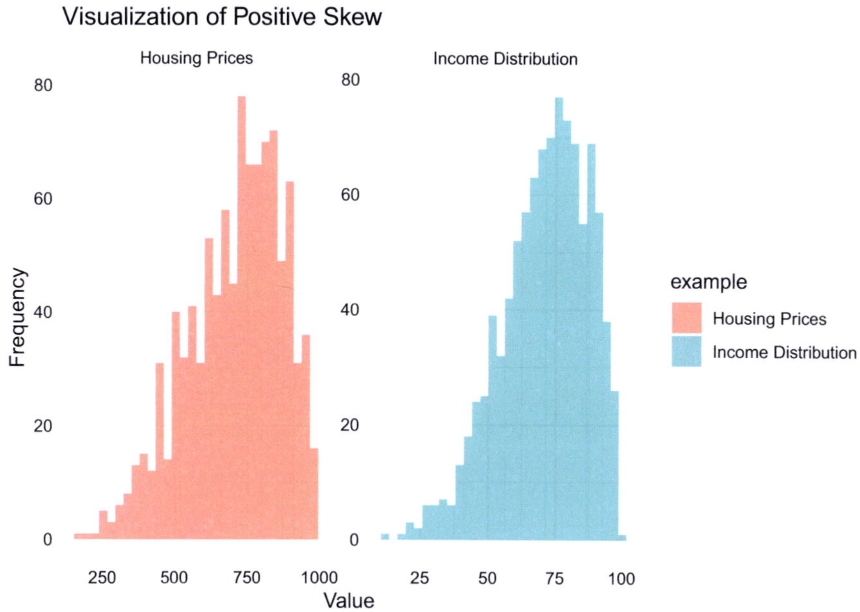

Fig. 2.1 Visualization of positie skew

```
# Simulate data for negative skew
set.seed(123)
negative_skew_test <-
    10 - rbeta(1000, 5, 2) * 10  # Easy test scores example
negative_skew_retirement <-
    80 - rbeta(1000, 5, 2) * 20  # Retirement age example

# Combine data
data_negative_skew <- data.frame(
    value = c(negative_skew_test, negative_skew_retirement),
    example = c(rep("Easy Test Scores", 1000),
                rep("Retirement Age", 1000))
)

# Plot negative skew
ggplot(data_negative_skew, aes(x = value, fill = example)) +
    geom_histogram(bins = 30,
                   alpha = 0.7,
                   position = "identity") +
    facet_wrap( ~ example, scales = "free") +
    labs(title = "Visualization of Negative Skew",
         x = "Value",
         y = "Frequency") +
    theme_minimal()
```

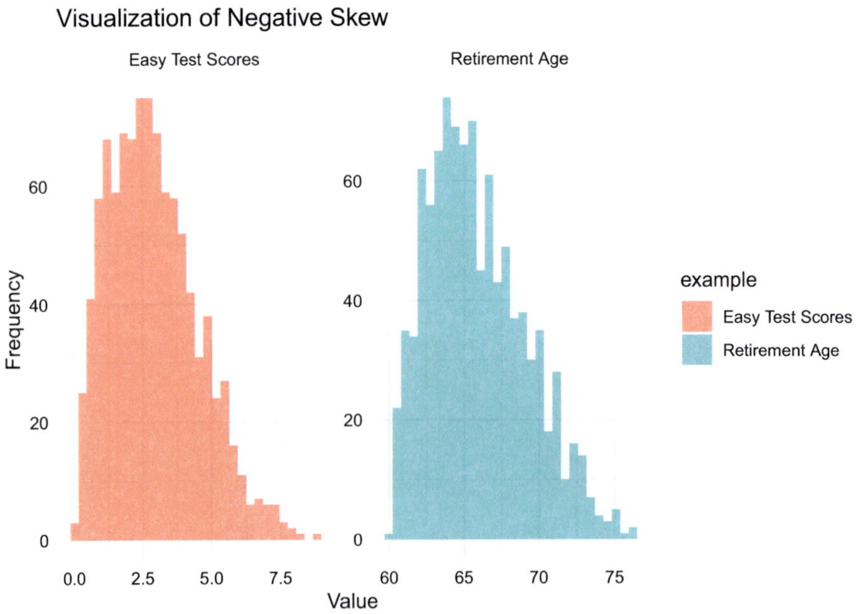

Fig. 2.2 Visualization of negative skew

- In the *easy test scores* example, most students perform well, but a few low scores stretch the left tail.
- In the *retirement age* example, most people retire around the same age, but a small number of individuals retire very early, stretching the left tail.

2.2.7 Kurtosis

Kurtosis measures the "peakedness" or heaviness of the tails:

- **High kurtosis**: Tall, sharp peak with heavy tails.
  - **Example**: Financial market returns during a crisis (extreme losses or gains).
- **Low kurtosis**: Flatter peak with thinner tails.
  - **Example**: Human height distribution (fewer extreme deviations from the mean) (Fig. 2.3).

```
# Simulate data for kurtosis
low_kurtosis <- runif(1000, 0, 10)
high_kurtosis <- c(rnorm(900, 5, 1), rnorm(100, 5, 5))

# Combine data
data_kurtosis <- data.frame(
  value = c(low_kurtosis, high_kurtosis),
  kurtosis_type = c(rep("Low Kurtosis (Height Distribution)", 1000),
                    rep("High Kurtosis (Market Returns)", 1000))
)

# Plot kurtosis
ggplot(data_kurtosis, aes(x = value, fill = kurtosis_type)) +
  geom_histogram(bins = 30, alpha = 0.7, position = "identity") +
  facet_wrap(~kurtosis_type) +
  labs(
    title = "Visualization of Kurtosis",
    x = "Value",
    y = "Frequency"
  ) +
  theme_minimal()
```

- The left panel shows *low kurtosis*, similar to the distribution of human height, which has a flatter peak and thinner tails.
- The right panel shows *high kurtosis*, reflecting financial market returns, where there are more extreme outliers in gains or losses.

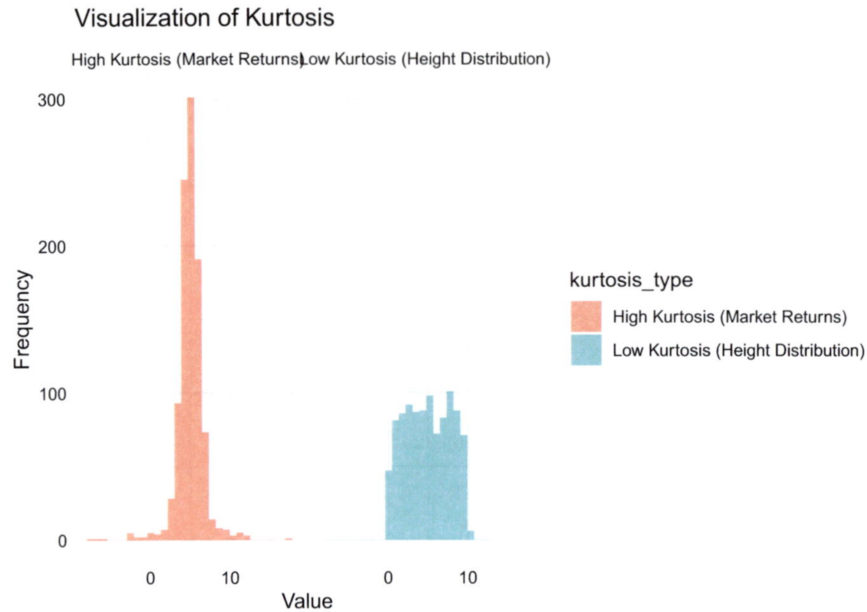

Fig. 2.3 Visualization of kurtosis

2.2.7.1 Conditional Moments

For a random variable Y given $X = x$:

1. **Expected Value**:

$$E[Y|X = x] = \begin{cases} \sum_y y f_Y(y|x) & \text{for discrete RV,} \\ \int_y y f_Y(y|x)\, dy & \text{for continuous RV.} \end{cases}$$

2. **Variance**:

$$\mathrm{Var}(Y|X = x) = \begin{cases} \sum_y (y - E[Y|X = x])^2 f_Y(y|x) & \text{for discrete RV,} \\ \int_y (y - E[Y|X = x])^2 f_Y(y|x)\, dy & \text{for continuous RV.} \end{cases}$$

2.2.7.2 Multivariate Moments

1. **Expected Value**:

$$E\begin{bmatrix} X \\ Y \end{bmatrix} = \begin{bmatrix} E[X] \\ E[Y] \end{bmatrix} = \begin{bmatrix} \mu_X \\ \mu_Y \end{bmatrix}.$$

2. **Variance-Covariance Matrix**:

$$\begin{aligned} \text{Var}\begin{bmatrix} X \\ Y \end{bmatrix} &= \begin{bmatrix} \text{Var}(X) & \text{Cov}(X, Y) \\ \text{Cov}(X, Y) & \text{Var}(Y) \end{bmatrix} \\ &= \begin{bmatrix} E[(X-\mu_X)^2] & E[(X-\mu_X)(Y-\mu_Y)] \\ E[(X-\mu_X)(Y-\mu_Y)] & E[(Y-\mu_Y)^2] \end{bmatrix}. \end{aligned}$$

2.2.7.3 Properties of Moments

1. $E[aX + bY + c] = aE[X] + bE[Y] + c$.
2. $\text{Var}(aX + bY + c) = a^2\text{Var}(X) + b^2\text{Var}(Y) + 2ab\text{Cov}(X, Y)$.
3. $\text{Cov}(aX + bY, cX + dY) = ac\text{Var}(X) + bd\text{Var}(Y) + (ad + bc)\text{Cov}(X, Y)$.
4. Correlation: $\rho_{XY} = \frac{\text{Cov}(X,Y)}{\sigma_X \sigma_Y}$.

2.2.8 Distributions

2.2.8.1 Conditional Distributions

$$f_{X|Y}(x|y) = \frac{f(x, y)}{f_Y(y)}$$

If X and Y are independent:

$$f_{X|Y}(x|y) = f_X(x)$$

2.2.8.2 Discrete Distributions

2.2.8.2.1 Bernoulli Distribution

A random variable X follows a Bernoulli distribution, denoted as $X \sim \text{Bernoulli}(p)$, if it represents a single trial with:

- Success probability p
- Failure probability $q = 1 - p$

Density Function:

$$f(x) = p^x (1-p)^{1-x}, \quad x \in \{0, 1\}$$

CDF: Use table or manual computation (Fig. 2.4).
PDF

```
hist(
    mc2d::rbern(1000, prob = 0.5),
    main = "Histogram of Bernoulli Distribution",
    xlab = "Value",
    ylab = "Frequency"
)
```

Mean

$$\mu = E[X] = p$$

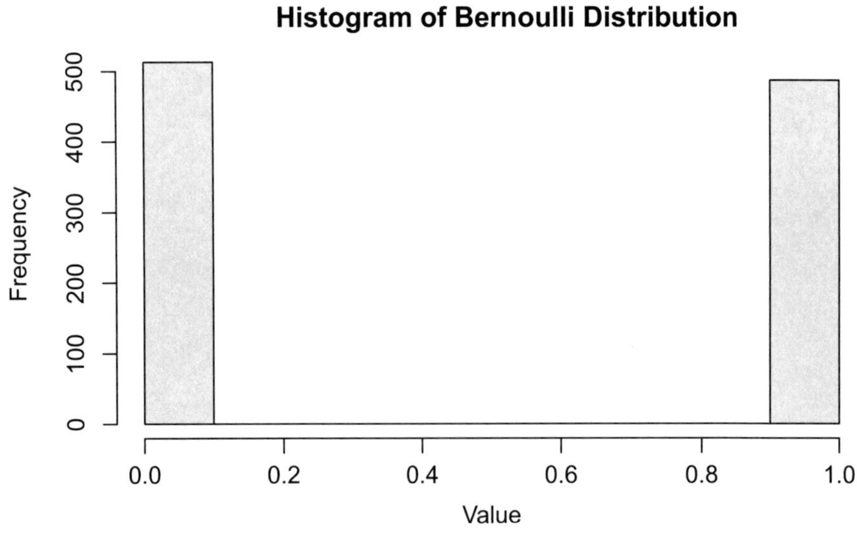

Fig. 2.4 Histogram of Bernoulli distribution

Variance

$$\sigma^2 = \mathrm{Var}(X) = p(1-p)$$

2.2.8.2.2 Binomial Distribution

$X \sim B(n, p)$ is the number of successes in n independent Bernoulli trials, where:

- n is the number of trials.
- p is the success probability.
- The trials are identical and independent, and probability of success (p) and probability of failure ($q = 1 - p$) remain the same for all trials (Fig. 2.5).

Density Function:

$$f(x) = \binom{n}{x} p^x (1-p)^{n-x}, \quad x = 0, 1, \ldots, n$$

PDF

```
hist(
    rbinom(1000, size = 100, prob = 0.5),
    main = "Histogram of Binomial Distribution",
    xlab = "Value",
    ylab = "Frequency"
)
```

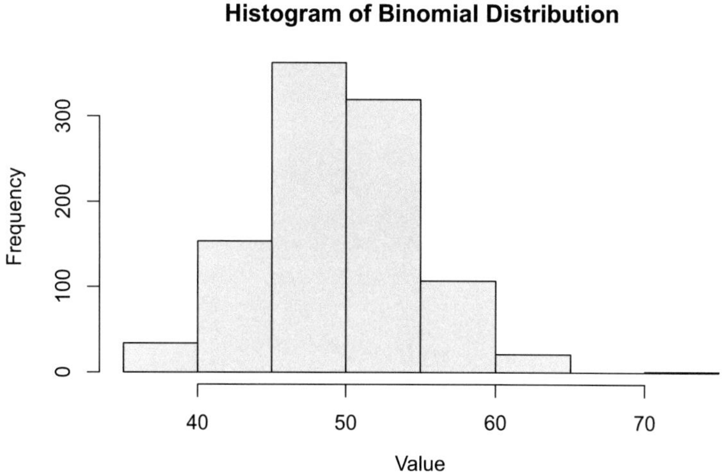

Fig. 2.5 Histogram of binomial distribution

MGF:

$$m_X(t) = (1 - p + pe^t)^n$$

Mean:

$$\mu = np$$

Variance:

$$\sigma^2 = np(1 - p)$$

2.2.8.2.3 Poisson Distribution

$X \sim \text{Poisson}(\lambda)$ models the number of occurrences of an event in a fixed interval, with average rate λ:

- Arises with Poisson process, which involves observing discrete events in a continuous "interval" of time, length, or space.
- The random variable X is the number of occurrences of the event within an interval of s units.
- The parameter λ is the average number of occurrences of the event in question per measurement unit. For the distribution, we use the parameter $k = \lambda s$ (Fig. 2.6).

Density Function:

$$f(x) = \frac{e^{-k}k^x}{x!}, \quad x = 0, 1, 2, \ldots$$

CDF
PDF

```
hist(rpois(1000, lambda = 5),
     main = "Histogram of Poisson Distribution",
     xlab = "Value",
     ylab = "Frequency")
```

MGF

$$m_X(t) = e^{k(e^t - 1)}$$

Mean:

$$\mu = E(X) = k$$

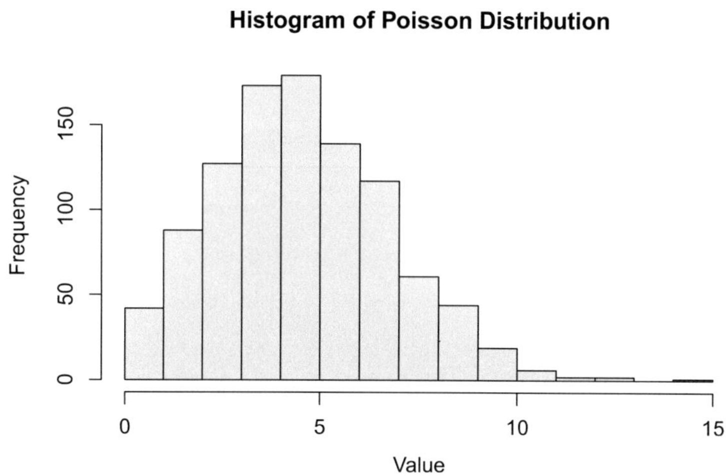

Fig. 2.6 Histogram of Poisson distribution

Variance:

$$\sigma^2 = Var(X) = k$$

2.2.8.2.4 Geometric Distribution

$X \sim \text{G}(p)$ models the number of trials needed to obtain the first success, with:

- p: probability of success
- $q = 1 - p$: probability of failure
- The experiment consists of a series of trails. The outcome of each trial can be classed as being either a "success" (s) or "failure" (f) (i.e., Bernoulli trial).
- The trials are identical and independent in the sense that the outcome of one trial has no effect on the outcome of any other (i.e., lack of memory—momerylessness). The probability of success (p) and probability of failure ($q = 1 - p$) remain the same from trial to trial.

Density Function:

$$f(x) = p(1-p)^{x-1}, \quad x = 1, 2, \ldots$$

CDF:

$$F(x) = 1 - (1-p)^x$$

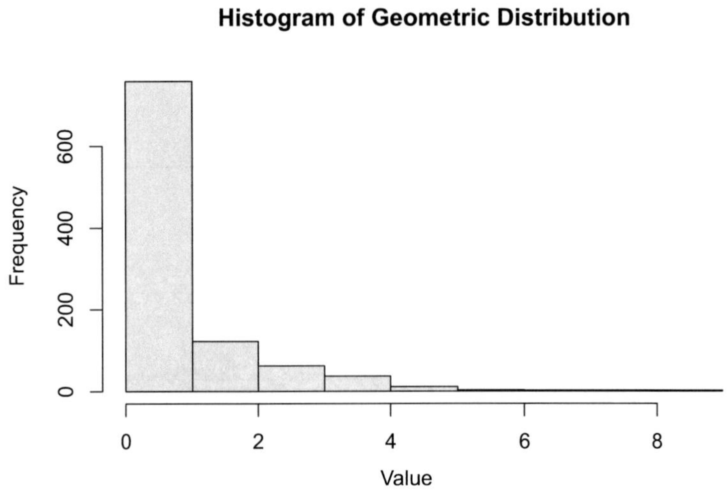

Fig. 2.7 Histogram of geometric distribution

PDF: (Fig. 2.7)

```
hist(rgeom(1000, prob = 0.5),
     main = "Histogram of Geometric Distribution",
     xlab = "Value",
     ylab = "Frequency")
```

MGF:

$$m_X(t) = \frac{pe^t}{1 - (1 - p)e^t}, \quad t < -\ln(1 - p)$$

Mean:

$$\mu = \frac{1}{p}$$

Variance:

$$\sigma^2 = \frac{1 - p}{p^2}$$

2.2.8.2.5 Hypergeometric Distribution

$X \sim \mathrm{H}(N, r, n)$ models the number of successes in a sample of size n drawn without replacement from a population of size N, where:

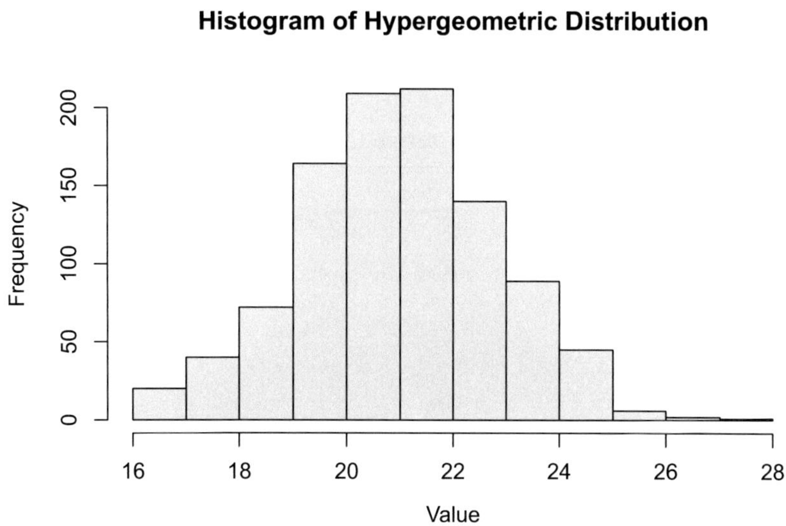

Fig. 2.8 Histogram of hypergeometric distribution

- r objects have the trait of interest.
- $N - r$ do not have the trait.

Density Function:

$$f(x) = \frac{\binom{r}{x}\binom{N-r}{n-x}}{\binom{N}{n}}, \quad \max(0, n - (N - r)) \leq x \leq \min(n, r)$$

PDF: (Fig. 2.8)

```
hist(
    rhyper(1000, m = 50, n = 20, k = 30),
    main = "Histogram of Hypergeometric Distribution",
    xlab = "Value",
    ylab = "Frequency"
)
```

Mean:

$$\mu = E[X] = \frac{nr}{N}$$

Variance:

$$\sigma^2 = \text{Var}(X) = n\frac{r}{N}\frac{N-r}{N}\frac{N-n}{N-1}$$

Note For large N (when $\frac{n}{N} \leq 0.05$), the hypergeometric distribution can be approximated by a binomial distribution with $p = \frac{r}{N}$.

2.2.8.3 Continuous Distributions

2.2.8.3.1 Uniform Distribution

Defined over an interval (a, b), where the probabilities are "equally likely" for subintervals of equal length.

Density Function:

$$f(x) = \frac{1}{b-a}, \quad a < x < b$$

CDF:

$$F(x) = \begin{cases} 0 & \text{if } x < a \\ \frac{x-a}{b-a} & a \leq x \leq b \\ 1 & \text{if } x > b \end{cases}$$

PDF: (Fig. 2.9)

```
hist(
    runif(1000, min = 0, max = 1),
    main = "Histogram of Uniform Distribution",
    xlab = "Value",
    ylab = "Frequency"
)
```

MGF:

$$m_X(t) = \begin{cases} \frac{e^{tb}-e^{ta}}{t(b-a)} & \text{if } t \neq 0 \\ 1 & \text{if } t = 0 \end{cases}$$

Mean:

$$\mu = E[X] = \frac{a+b}{2}$$

Variance:

$$\sigma^2 = \text{Var}(X) = \frac{(b-a)^2}{12}$$

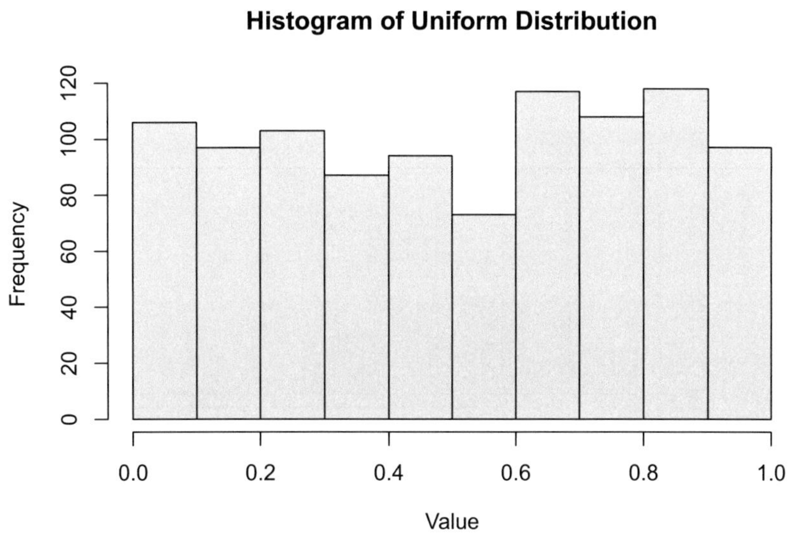

Fig. 2.9 Histogram of uniform distribution

2.2.8.3.2 Gamma Distribution

The gamma distribution is used to define the exponential and χ^2 distributions.

The *gamma function* is defined as:

$$\Gamma(\alpha) = \int_0^{\infty} z^{\alpha-1} e^{-z} dz, \quad \alpha > 0$$

Properties of the Gamma Function:

- $\Gamma(1) = 1$.
- For $\alpha > 1$, $\Gamma(\alpha) = (\alpha - 1)\Gamma(\alpha - 1)$.
- If n is an integer and $n > 1$, then $\Gamma(n) = (n - 1)!$.

Density Function:

$$f(x) = \frac{1}{\Gamma(\alpha)\beta^{\alpha}} x^{\alpha-1} e^{-x/\beta}, \quad x > 0$$

CDF (for $\alpha = n$, and $x > 0$ a positive integer):

$$F(x, n, \beta) = 1 - \sum_{k=0}^{n-1} \frac{(\frac{x}{\beta})^k e^{-x/\beta}}{k!}$$

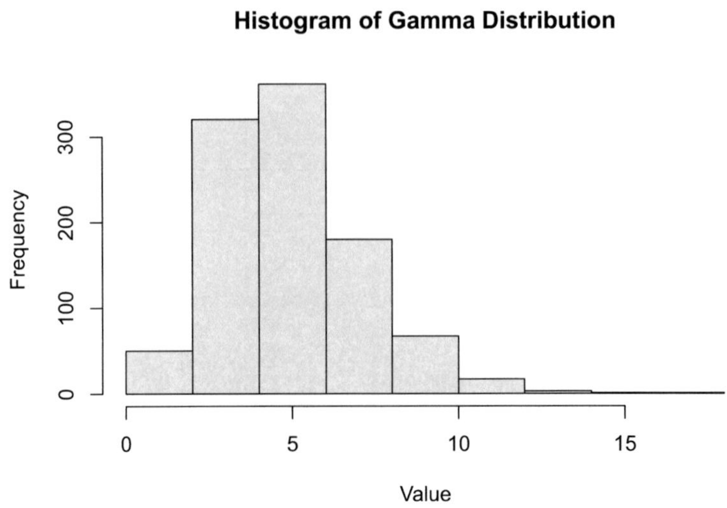

Fig. 2.10 Histogram of gamma distribution

PDF: (Fig. 2.10)

```
hist(
    rgamma(n = 1000, shape = 5, rate = 1),
    main = "Histogram of Gamma Distribution",
    xlab = "Value",
    ylab = "Frequency"
)
```

MGF:

$$m_X(t) = (1 - \beta t)^{-\alpha}, \quad t < \frac{1}{\beta}$$

Mean:

$$\mu = E[X] = \alpha\beta$$

Variance:

$$\sigma^2 = \mathrm{Var}(X) = \alpha\beta^2$$

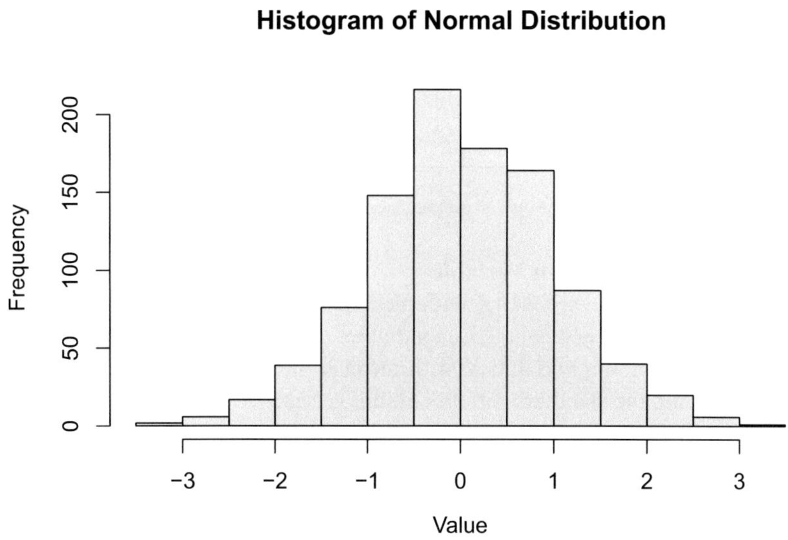

Fig. 2.11 Histogram of normal distribution

2.2.8.3.3 Normal Distribution

The normal distribution, denoted as $N(\mu, \sigma^2)$, is symmetric and bell-shaped with parameters μ (mean) and σ^2 (variance). It is also known as the Gaussian distribution.

Density Function:

$$f(x) = \frac{1}{\sigma\sqrt{2\pi}} e^{-\frac{1}{2}\left(\frac{x-\mu}{\sigma}\right)^2}, \quad -\infty < x < \infty,\ \sigma > 0$$

CDF: Use table or numerical methods.
PDF (Fig. 2.11)

```
hist(
    rnorm(1000, mean = 0, sd = 1),
    main = "Histogram of Normal Distribution",
    xlab = "Value",
    ylab = "Frequency"
)
```

MGF:

$$m_X(t) = e^{\mu t + \frac{\sigma^2 t^2}{2}}$$

Mean:

$$\mu = E[X]$$

Variance:

$$\sigma^2 = \mathrm{Var}(X)$$

Standard Normal Random Variable

- The normal random variable Z with mean $\mu = 0$ and standard deviation $\sigma = 1$ is called a standard normal random variable.
- Any normal random variable X with mean μ and standard deviation σ can be converted into the standard normal random variable Z:

$$Z = \frac{X - \mu}{\sigma}.$$

Normal Approximation to the Binomial Distribution
Let X be binomial with parameters n and p. For large n:

- If $p \leq 0.5$ and $np > 5$, or
- If $p > 0.5$ and $n(1 - p) > 5$

X is approximately normally distributed with mean $\mu = np$ and standard deviation $\sigma = \sqrt{np(1 - p)}$.

When using the normal approximation, add or subtract 0.5 as needed for the continuity correction (Table 2.4).

Discrete Approximate Normal (Corrected)

Table 2.4 Continuity correction for normal approximation to discrete probabilities

Discrete	Approximate normal (corrected)
$P(X = c)$	$P(c - 0.5 < Y < c + 0.5)$
$P(X < c)$	$P(Y < c - 0.5)$
$P(X \leq c)$	$P(Y < c + 0.5)$
$P(X > c)$	$P(Y > c + 0.5)$
$P(X \geq c)$	$P(Y > c - 0.5)$

Normal Probability Rule
If X is normally distributed with parameters μ and σ, then:

- $P(-\sigma < X - \mu < \sigma) \approx .68$.
- $P(-2\sigma < X - \mu < 2\sigma) \approx .95$.
- $P(-3\sigma < X - \mu < 3\sigma) \approx .997$.

2.2.8.3.4 Logistic Distribution

The logistic distribution is a continuous probability distribution commonly used in logistic regression and other types of statistical modeling. It resembles the normal distribution, but has heavier tails, allowing for more extreme values. The logistic distribution is symmetric around μ. Its heavier tails make it useful for modeling outcomes with occasional extreme values.

Density Function:

$$f(x; \mu, s) = \frac{e^{-(x-\mu)/s}}{s\left(1 + e^{-(x-\mu)/s}\right)^2}, \quad -\infty < x < \infty$$

where μ is the location parameter (mean) and $s > 0$ is the scale parameter.

CDF:

$$F(x; \mu, s) = \frac{1}{1 + e^{-(x-\mu)/s}}, \quad -\infty < x < \infty$$

PDF: (Fig. 2.12)

```
hist(
    rlogis(1000, location = 0, scale = 1),
    main = "Histogram of Logistic Distribution",
    xlab = "Value",
    ylab = "Frequency"
)
```

MGF:
The MGF of the logistic distribution does not exist because its expected value diverges for most t.

Mean:

$$\mu = E[X] = \mu$$

Variance:

$$\sigma^2 = \text{Var}(X) = \frac{\pi^2 s^2}{3}$$

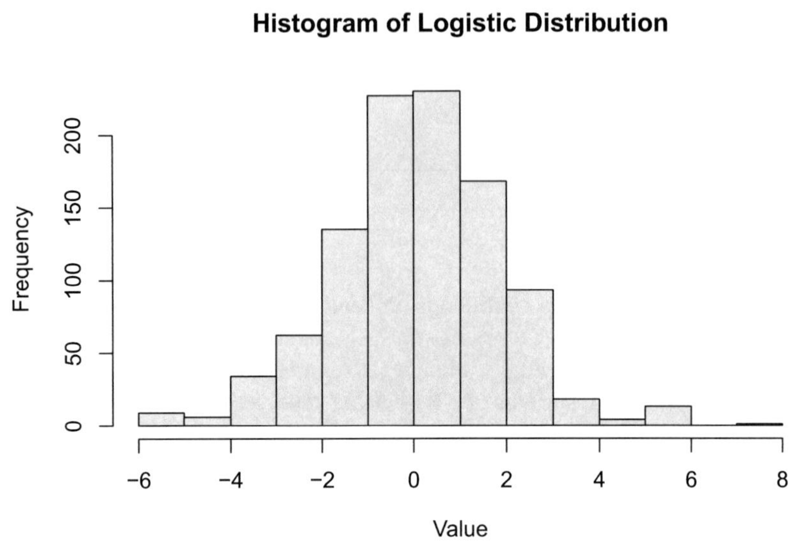

Fig. 2.12 Histogram of logistic distribution

2.2.8.3.5 Laplace Distribution

The Laplace distribution, also known as the double exponential distribution, is a continuous probability distribution often used in economics, finance, and engineering. It is characterized by a peak at its mean and heavier tails compared to the normal distribution:

- The Laplace distribution is symmetric around μ.
- It has heavier tails than the normal distribution, making it suitable for modeling data with more extreme outliers.

Density Function:

$$f(x; \mu, b) = \frac{1}{2b} e^{-|x-\mu|/b}, \quad -\infty < x < \infty$$

where μ is the location parameter (mean) and $b > 0$ is the scale parameter.

CDF:

$$F(x; \mu, b) = \begin{cases} \frac{1}{2} e^{(x-\mu)/b} & \text{if } x < \mu \\ 1 - \frac{1}{2} e^{-(x-\mu)/b} & \text{if } x \geq \mu \end{cases}$$

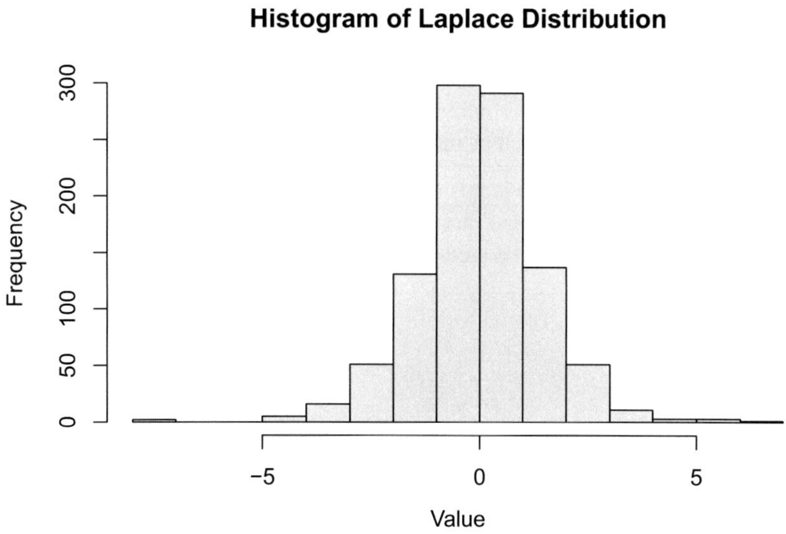

Fig. 2.13 Histogram of Laplace distribution

PDF: (Fig. 2.13)

```
hist(
    VGAM::rlaplace(1000, location = 0, scale = 1),
    main = "Histogram of Laplace Distribution",
    xlab = "Value",
    ylab = "Frequency"
)
```

MGF:

$$m_X(t) = \frac{e^{\mu t}}{1 - b^2 t^2}, \quad |t| < \frac{1}{b}$$

Mean:

$$\mu = E[X] = \mu$$

Variance:

$$\sigma^2 = \text{Var}(X) = 2b^2$$

2.2.8.3.6 Lognormal Distribution

The lognormal distribution is a continuous probability distribution of a random variable whose logarithm is normally distributed. It is often used to model variables that are positively skewed, such as income or biological measurements:

- The lognormal distribution is positively skewed.
- It is useful for modeling data that cannot take negative values and is often used in finance and environmental studies.

Density Function:

$$f(x; \mu, \sigma) = \frac{1}{x\sigma\sqrt{2\pi}} e^{-(\ln(x)-\mu)^2/(2\sigma^2)}, \quad x > 0$$

where μ is the mean of the underlying normal distribution and $\sigma > 0$ is the standard deviation.

CDF:
The cumulative distribution function of the lognormal distribution is given by

$$F(x; \mu, \sigma) = \frac{1}{2}\left[1 + \text{erf}\left(\frac{\ln(x) - \mu}{\sigma\sqrt{2}}\right)\right], \quad x > 0$$

PDF: (Fig. 2.14)

```
hist(
    rlnorm(1000, meanlog = 0, sdlog = 1),
    main = "Histogram of Lognormal Distribution",
    xlab = "Value",
    ylab = "Frequency"
)
```

MGF:
The moment generating function (MGF) of the lognormal distribution does not exist in a simple closed form.
Mean:

$$E[X] = e^{\mu+\sigma^2/2}$$

Variance:

$$\sigma^2 = \text{Var}(X) = \left(e^{\sigma^2} - 1\right) e^{2\mu+\sigma^2}$$

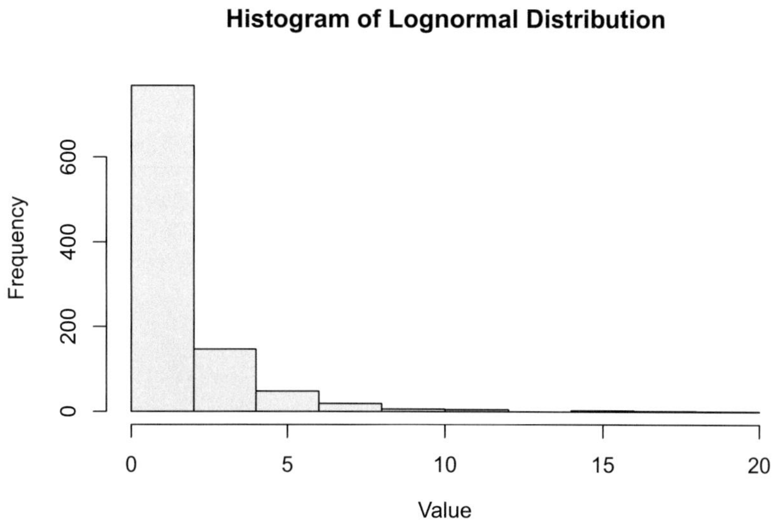

Fig. 2.14 Histogram of lognormal distribution

2.2.8.3.7 Exponential Distribution

The exponential distribution, denoted as Exp(λ), is a special case of the gamma distribution with $\alpha = 1$:

- It is commonly used to model the time between independent events that occur at a constant rate. It is often applied in reliability analysis and queuing theory.
- The exponential distribution is memoryless, meaning the probability of an event occurring in the future is independent of the past.
- It is commonly used to model waiting times, such as the time until the next customer arrives or the time until a radioactive particle decays.

Density Function:

$$f(x) = \frac{1}{\beta} e^{-x/\beta}, \quad x, \beta > 0$$

CDF:

$$F(x) = \begin{cases} 0 & \text{if } x \leq 0 \\ 1 - e^{-x/\beta} & \text{if } x > 0 \end{cases}$$

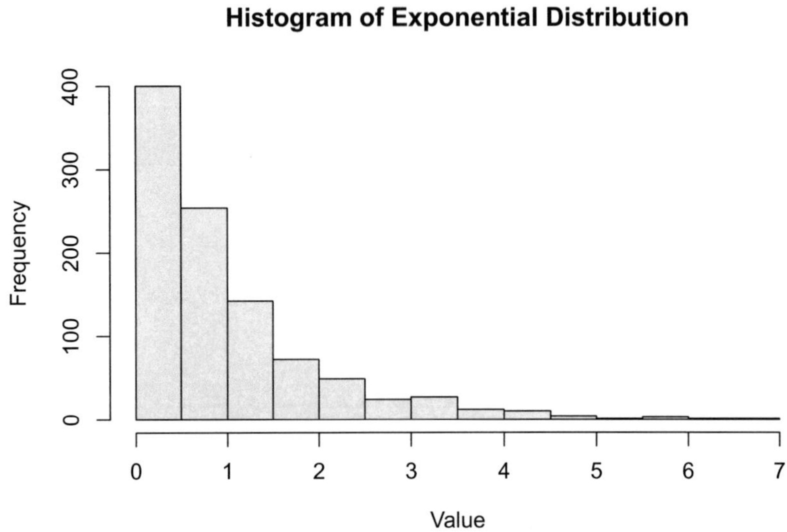

Fig. 2.15 Histogram of exponential distribution

PDF: (Fig. 2.15)

```
hist(rexp(n = 1000, rate = 1),
     main = "Histogram of Exponential Distribution",
     xlab = "Value",
     ylab = "Frequency")
```

MGF:

$$m_X(t) = (1 - \beta t)^{-1}, \quad t < 1/\beta$$

Mean:

$$\mu = E[X] = \beta$$

Variance:

$$\sigma^2 = \text{Var}(X) = \beta^2$$

2.2.8.3.8 Chi-Squared Distribution

The chi-squared distribution is a continuous probability distribution commonly used in statistical inference, particularly in hypothesis testing and construction of confidence intervals for variance. It is also used in goodness-of-fit tests:

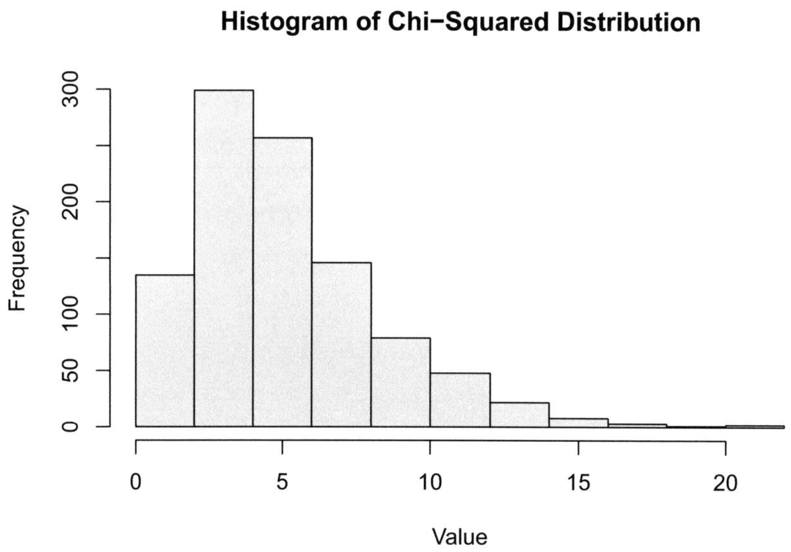

Fig. 2.16 Histogram of chi-squared distribution

- The chi-squared distribution is defined only for positive values.
- It is often used to model the distribution of the sum of the squares of k independent standard normal random variables.

Density Function:

$$f(x; k) = \frac{1}{2^{k/2}\Gamma(k/2)} x^{k/2-1} e^{-x/2}, \quad x \geq 0$$

where k is the degrees of freedom and Γ is the gamma function.

CDF

The cumulative distribution function of the chi-squared distribution is given by

$$F(x; k) = \frac{\gamma(k/2, x/2)}{\Gamma(k/2)}, \quad x \geq 0$$

where γ is the lower incomplete gamma function.

PDF: (Fig. 2.16)

```
hist(
    rchisq(1000, df = 5),
    main = "Histogram of Chi-Squared Distribution",
    xlab = "Value",
    ylab = "Frequency"
)
```

MGF:

$$m_X(t) = (1 - 2t)^{-k/2}, \quad t < \frac{1}{2}$$

Mean:

$$E[X] = k$$

Variance:

$$\sigma^2 = \text{Var}(X) = 2k$$

2.2.8.3.9 Student's T-Distribution

The *Student's t-distribution* is named after *William Sealy Gosset*, a statistician at Guinness Brewery in the early 20th century. Gosset developed the t-distribution to address small sample problems in quality control. Since Guinness prohibited employees from publishing under their names, Gosset used the pseudonym *"Student"* when he published his work in 1908 (Student, 1908). The name has stuck ever since, honoring his contribution to statistics.

The Student's t-distribution, denoted as $T(v)$, is defined by

$$T = \frac{Z}{\sqrt{\chi_v^2/v}}$$

where Z is a standard normal random variable and χ_v^2 follows a chi-squared distribution with v degrees of freedom.

The Student's t-distribution is a continuous probability distribution used in statistical inference, particularly for estimating population parameters when the sample size is small and/or the population variance is unknown. It is similar to the normal distribution, but has heavier tails, which makes it more robust for small sample sizes:

- The Student's t-distribution is symmetric around 0.
- It has heavier tails than the normal distribution, making it useful for dealing with outliers or small sample sizes.

Density Function:

$$f(x; v) = \frac{\Gamma((v+1)/2)}{\sqrt{v\pi}\,\Gamma(v/2)} \left(1 + \frac{x^2}{v}\right)^{-(v+1)/2}$$

where v is the degrees of freedom and $\Gamma(x)$ is the gamma function.

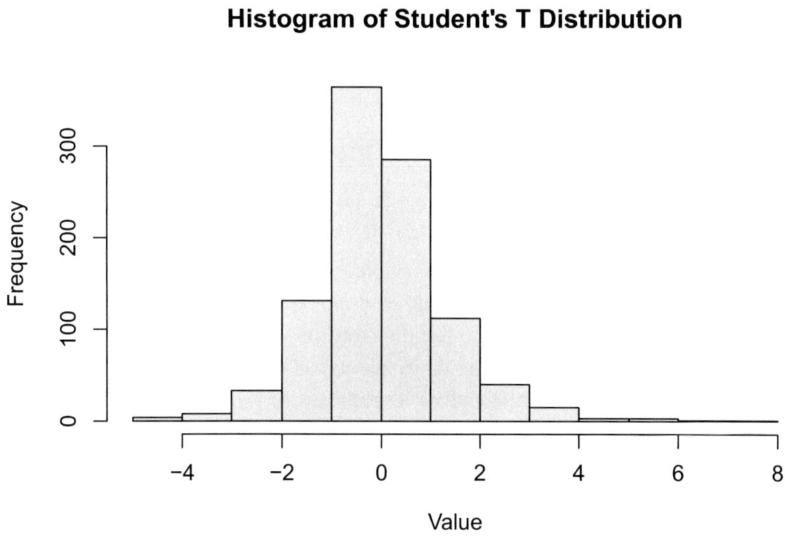

Fig. 2.17 Histogram of Student's t-distribution

CDF

The cumulative distribution function of the Student's t-distribution is more complex and typically evaluated using numerical methods.

PDF (Fig. 2.17)

```
hist(
    rt(1000, df = 5),
    main = "Histogram of Student's T Distribution",
    xlab = "Value",
    ylab = "Frequency"
)
```

MGF:

The moment generating function (MGF) of the Student's t-distribution does not exist in a simple closed form.

Mean:

For $v > 1$:

$$E[X] = 0$$

Variance:

For $v > 2$:

$$\sigma^2 = \text{Var}(X) = \frac{v}{v - 2}$$

2.2.8.3.10 Noncentral t-distribution

The noncentral t-distribution, denoted as $T(v, \lambda)$, is a generalization of the Student's t-distribution. It is defined as

$$T = \frac{Z + \lambda}{\sqrt{\chi_v^2 / v}}$$

where Z is a standard normal random variable, χ_v^2 follows a chi-squared distribution with v degrees of freedom, and λ is the *non-centrality parameter*. This additional parameter introduces asymmetry to the distribution.

The noncentral t-distribution arises in scenarios where the null hypothesis does not hold, such as under the alternative hypothesis in hypothesis testing. The non-centrality parameter λ represents the degree to which the mean deviates from zero:

- For $\lambda = 0$, the noncentral t-distribution reduces to the Student's t-distribution.
- The distribution is skewed for $\lambda \neq 0$, with the skewness increasing as λ grows.

Density Function:
The density function of the noncentral t-distribution is more complex and depends on v and λ. It can be expressed in terms of an infinite sum:

$$f(x; v, \lambda) = \sum_{k=0}^{\infty} \frac{e^{-\lambda^2/2} (\lambda^2/2)^k}{k!} \cdot \frac{\Gamma((v+k+1)/2)}{\sqrt{v\pi}\,\Gamma((v+k)/2)} \left(1 + \frac{x^2}{v}\right)^{-(v+k+1)/2}.$$

PDF: (Fig. 2.18)

```
n <- 100  # Number of samples
df <- 5    # Degrees of freedom
lambda <- 2  # Non-centrality parameter


hist(
  rt(n, df = df, ncp = lambda),
  main = "Histogram of Non-central T Distribution",
  xlab = "Value",
  ylab = "Frequency"
)
```

CDF:
The cumulative distribution function of the noncentral t-distribution is typically computed using numerical methods due to its complexity.
Mean:
For $v > 1$:

$$E[T] = \lambda \sqrt{\frac{v}{2}} \cdot \frac{\Gamma((v-1)/2)}{\Gamma(v/2)}$$

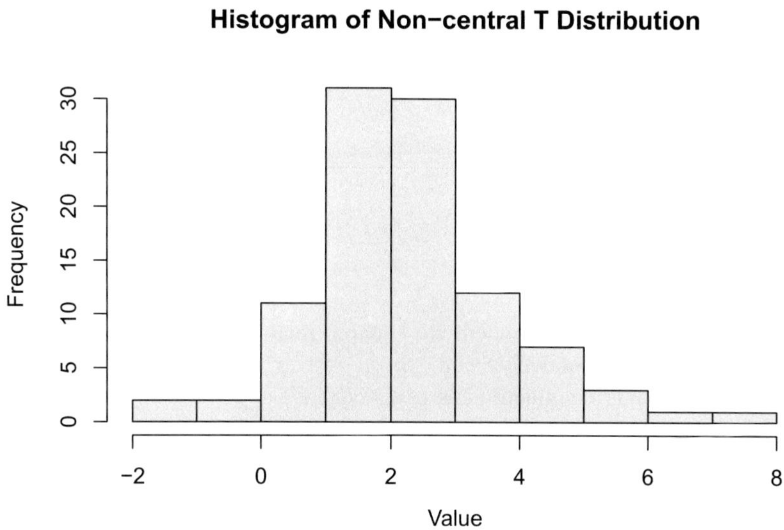

Fig. 2.18 Histogram of noncentral t-distribution

Variance:
For $v > 2$ (Table 2.5):

$$\mathrm{Var}(T) = \frac{v}{v-2} + \lambda^2$$

While the Student's t-distribution is used for standard hypothesis testing and confidence intervals, the noncentral t-distribution finds its applications in scenarios involving non-null hypotheses, such as power and sample size calculations.

Table 2.5 Comparison between Student's T and noncentral T

Feature	Student's t-distribution	Noncentral t-distribution
Definition	$T = \frac{Z}{\sqrt{\chi_v^2/v}}$	$T = \frac{Z+\lambda}{\sqrt{\chi_v^2/v}}$
Centered at	0	λ
Symmetry	Symmetric	Skewed for $\lambda \neq 0$
Parameters	Degrees of freedom (v)	v and λ
Shape as $v \to \infty$ (df $\to \infty$)	Normal(0, 1)	Normal(λ, 1)
Applications	Hypothesis testing under null	Power analysis, alternative testing

2.2.8.3.11 F-Distribution

The F-distribution, denoted as $F(d_1, d_2)$, is strictly positive and used to compare variances.

Definition

$$F = \frac{\chi^2_{d_1}/d_1}{\chi^2_{d_2}/d_2}$$

where $\chi^2_{d_1}$ and $\chi^2_{d_2}$ are independent chi-squared random variables with degrees of freedom d_1 and d_2, respectively.

The distribution is asymmetric and never negative.

The F-distribution arises frequently as the null distribution of a test statistic, especially in the context of comparing variances, such as in analysis of variance (ANOVA).

Density Function:

$$f(x; d_1, d_2) = \frac{\sqrt{\frac{(d_1 x)^{d_1} d_2^{d_2}}{(d_1 x + d_2)^{d_1+d_2}}}}{x B\left(\frac{d_1}{2}, \frac{d_2}{2}\right)}, \quad x > 0$$

where d_1 and d_2 are the degrees of freedom and B is the beta function.

CDF:

The cumulative distribution function of the F-distribution is typically evaluated using numerical methods.

PDF: (Fig. 2.19)

```
hist(
    rf(1000, df1 = 5, df2 = 2),
    main = "Histogram of F Distribution",
    xlab = "Value",
    ylab = "Frequency"
)
```

MGF:

The moment generating function (MGF) of the F-distribution does not exist in a simple closed form.

Mean:

For $d_2 > 2$:

$$E[X] = \frac{d_2}{d_2 - 2}$$

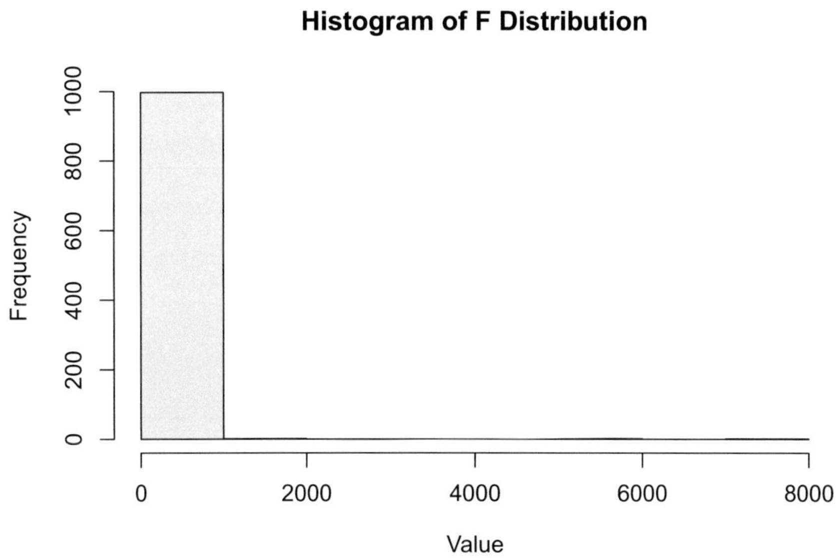

Fig. 2.19 Histogram of F-distribution

Variance:
For $d_2 > 4$:

$$\sigma^2 = \text{Var}(X) = \frac{2d_2^2(d_1 + d_2 - 2)}{d_1(d_2 - 2)^2(d_2 - 4)}$$

2.2.8.3.12 Cauchy Distribution

The Cauchy distribution is a continuous probability distribution that is often used in physics and has heavier tails than the normal distribution. It is notable because it does not have a finite mean or variance:

- The Cauchy distribution does not have a finite mean or variance.
- The central limit theorem and weak law of large numbers do not apply to the Cauchy distribution.

Density Function:

$$f(x; x_0, \gamma) = \frac{1}{\pi\gamma\left[1 + \left(\frac{x - x_0}{\gamma}\right)^2\right]}$$

where x_0 is the location parameter and $\gamma > 0$ is the scale parameter.

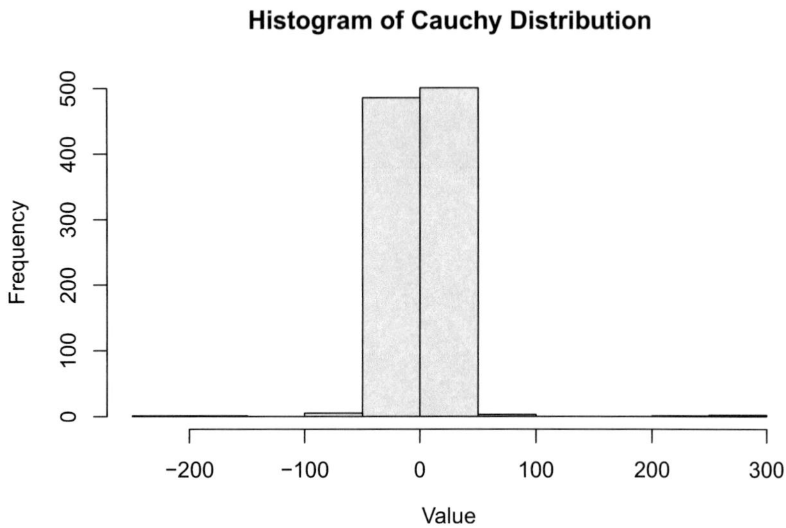

Fig. 2.20 Histogram of Cauchy distribution

CDF:
The cumulative distribution function of the Cauchy distribution is given by

$$F(x; x_0, \gamma) = \frac{1}{\pi} \arctan\left(\frac{x - x_0}{\gamma}\right) + \frac{1}{2}$$

PDF: (Fig. 2.20)

```
hist(
    rcauchy(1000, location = 0, scale = 1),
    main = "Histogram of Cauchy Distribution",
    xlab = "Value",
    ylab = "Frequency"
)
```

MGF:
The MGF of the Cauchy distribution does not exist.
Mean:
The mean of the Cauchy distribution is undefined.
Variance:
The variance of the Cauchy distribution is undefined.

2.2.8.3.13 Multivariate Normal Distribution

Let y be a p-dimensional multivariate normal (MVN) random variable with mean μ and variance-covariance matrix Σ. The density function of y is given by

$$f(\mathbf{y}) = \frac{1}{(2\pi)^{p/2}|\mathbf{\Sigma}|^{1/2}} \exp\left(-\frac{1}{2}(\mathbf{y}-\mu)'\Sigma^{-1}(\mathbf{y}-\mu)\right)$$

where $|\mathbf{\Sigma}|$ represents the determinant of the variance-covariance matrix Σ, and $\mathbf{y} \sim N_p(\boldsymbol{\mu}, \mathbf{\Sigma})$.

Properties:

- Let $\mathbf{A}_{r\times p}$ be a fixed matrix. Then $\mathbf{A}\mathbf{y} \sim N_r(\mathbf{A}\boldsymbol{\mu}, \mathbf{A}\mathbf{\Sigma}\mathbf{A}')$. Note that $r \leq p$, and all rows of $\mathbf{A}$ must be linearly independent to guarantee that $\mathbf{A}\mathbf{\Sigma}\mathbf{A}'$ is non-singular.
- Let $\mathbf{G}$ be a matrix such that $\mathbf{\Sigma}^{-1} = \mathbf{G}\mathbf{G}'$. Then $\mathbf{G}'\mathbf{y} \sim N_p(\mathbf{G}'\boldsymbol{\mu}, \mathbf{I})$ and $\mathbf{G}'(\mathbf{y}-\boldsymbol{\mu}) \sim N_p(\mathbf{0}, \mathbf{I})$.
- Any fixed linear combination of $y_1, \ldots, y_p$, say $\mathbf{c}'\mathbf{y}$, follows $\mathbf{c}'\mathbf{y} \sim N_1(\mathbf{c}'\boldsymbol{\mu}, \mathbf{c}'\mathbf{\Sigma}\mathbf{c})$.

Large Sample Properties

Suppose that $y_1, \ldots, y_n$ are a random sample from some population with mean μ and variance-covariance matrix Σ:

$$\mathbf{Y} \sim MVN(\boldsymbol{\mu}, \mathbf{\Sigma})$$

Then:

- $\bar{\mathbf{y}} = \frac{1}{n}\sum_{i=1}^{n} \mathbf{y}_i$ is a consistent estimator for $\boldsymbol{\mu}$.
- $\mathbf{S} = \frac{1}{n-1}\sum_{i=1}^{n}(\mathbf{y}_i - \bar{\mathbf{y}})(\mathbf{y}_i - \bar{\mathbf{y}})'$ is a consistent estimator for $\mathbf{\Sigma}$.
- **Multivariate Central Limit Theorem**: Similar to the univariate case, $\sqrt{n}(\bar{\mathbf{y}} - \mu) \sim N_p(\mathbf{0}, \mathbf{\Sigma})$ when n is large relative to p (e.g., $n \geq 25p$), which is equivalent to $\bar{\mathbf{y}} \sim N_p(\boldsymbol{\mu}, \mathbf{\Sigma}/\mathbf{n})$.
- **Wald's Theorem**: $n(\bar{\mathbf{y}} - \mu)'\mathbf{S}^{-1}(\bar{\mathbf{y}} - \mu) \sim \chi^2_{(p)}$ when n is large relative to p.

Density Function:

$$f(\mathbf{x}; \boldsymbol{\mu}, \mathbf{\Sigma}) = \frac{1}{(2\pi)^{k/2}|\mathbf{\Sigma}|^{1/2}} \exp\left(-\frac{1}{2}(\mathbf{x}-\boldsymbol{\mu})^T\mathbf{\Sigma}^{-1}(\mathbf{x}-\boldsymbol{\mu})\right)$$

where $\boldsymbol{\mu}$ is the mean vector, $\mathbf{\Sigma}$ is the covariance matrix, $\mathbf{x} \in \mathbb{R}^k$, and k is the number of variables.

CDF:

The cumulative distribution function of the multivariate normal distribution does not have a simple closed form and is typically evaluated using numerical methods.

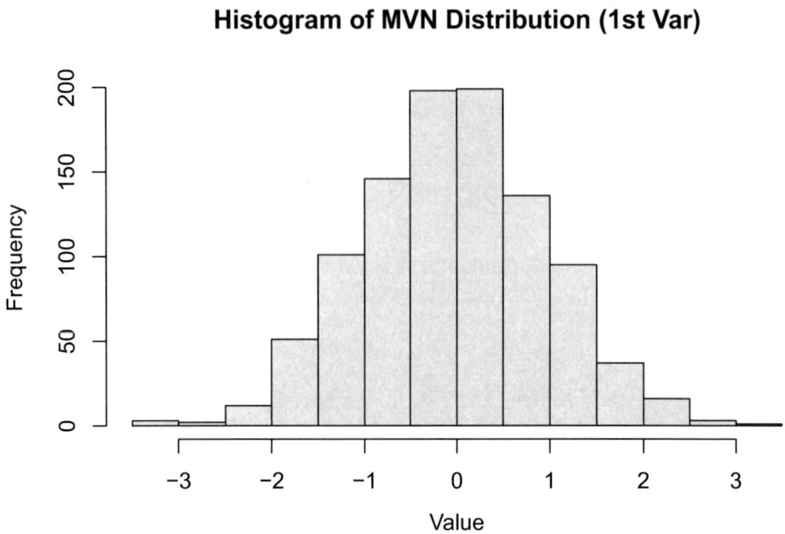

Fig. 2.21 Histogram of MVN distribution (1st Var)

PDF: (Fig. 2.21)

```
k <- 2
n <- 1000
mu <- c(0, 0)
sigma <- matrix(c(1, 0.5, 0.5, 1), nrow = k)
library(MASS)
hist(
    mvrnorm(n, mu = mu, Sigma = sigma)[,1],
    main = "Histogram of MVN Distribution (1st Var)",
    xlab = "Value",
    ylab = "Frequency"
)
```

MGF:

$$m_{\mathbf{X}}(\mathbf{t}) = \exp\left(\boldsymbol{\mu}^T\mathbf{t} + \frac{1}{2}\mathbf{t}^T\boldsymbol{\Sigma}\mathbf{t}\right)$$

Mean:

$$E[\mathbf{X}] = \boldsymbol{\mu}$$

Variance:

$$\text{Var}(\mathbf{X}) = \boldsymbol{\Sigma}$$

2.3 General Math

2.3.1 Number Sets

See Table 2.6

2.3.2 Summation Notation and Series

2.3.2.1 Chebyshev's Inequality

Let X be a random variable with mean μ and standard deviation σ. For any positive number k, Chebyshev's inequality states:

$$P(|X - \mu| \geq k\sigma) \leq \frac{1}{k^2}$$

This provides a probabilistic bound on the deviation of X from its mean and does not require X to follow a normal distribution.

2.3.2.2 Geometric Sum

For a geometric series of the form $\sum_{k=0}^{n-1} ar^k$, the sum is given by

$$\sum_{k=0}^{n-1} ar^k = a\frac{1 - r^n}{1 - r} \quad \text{where } r \neq 1$$

Table 2.6 Common mathematical set notations and examples

Notation	Denotes	Examples
$\emptyset$	Empty set	No members
$\mathbb{N}$	Natural numbers	$\{1, 2, \ldots\}$
$\mathbb{Z}$	Integers	$\{\ldots, -1, 0, 1, \ldots\}$
$\mathbb{Q}$	Rational numbers	Including fractions
$\mathbb{R}$	Real numbers	Including all finite decimals, irrational numbers
$\mathbb{C}$	Complex numbers	Including numbers of the form $a + bi$ where $i^2 = -1$

2.3.2.3 Infinite Geometric Series

When $|r| < 1$, the geometric series converges to

$$\sum_{k=0}^{\infty} ar^k = \frac{a}{1-r}$$

2.3.2.4 Binomial Theorem

The binomial expansion for $(x+y)^n$ is

$$(x+y)^n = \sum_{k=0}^{n} \binom{n}{k} x^{n-k} y^k \quad \text{where } n \geq 0$$

2.3.2.5 Binomial Series

For non-integer exponents α:

$$\sum_{k=0}^{\infty} \binom{\alpha}{k} x^k = (1+x)^\alpha \quad \text{where } |x| < 1$$

2.3.2.6 Telescoping Sum

A telescoping sum simplifies as intermediate terms cancel, leaving:

$$\sum_{a \leq k < b} \Delta F(k) = F(b) - F(a) \quad \text{where } a, b \in \mathbb{Z}, a \leq b$$

2.3.2.7 Vandermonde Convolution

The Vandermonde convolution identity is

$$\sum_{k=0}^{n} \binom{r}{k} \binom{s}{n-k} = \binom{r+s}{n} \quad \text{where } n \in \mathbb{Z}$$

2.3.2.8 Exponential Series

The exponential function e^x can be represented as

$$\sum_{k=0}^{\infty} \frac{x^k}{k!} = e^x \quad \text{where } x \in \mathbb{C}$$

2.3.2.9 Taylor Series

The Taylor series expansion for a function $f(x)$ about $x = a$ is

$$\sum_{k=0}^{\infty} \frac{f^{(k)}(a)}{k!}(x-a)^k = f(x)$$

For $a = 0$, this becomes the *Maclaurin series*.

2.3.2.10 Maclaurin Series for e^z

A special case of the Taylor series, the Maclaurin expansion for e^z, is

$$e^z = 1 + z + \frac{z^2}{2!} + \frac{z^3}{3!} + \cdots$$

2.3.2.11 Euler's Summation Formula

Euler's summation formula connects sums and integrals:

$$\sum_{a \le k < b} f(k) = \int_a^b f(x)\,dx + \sum_{k=1}^{m} \frac{B_k}{k!} \left[f^{(k-1)}(x) \right]_a^b + (-1)^{m+1} \int_a^b \frac{B_m(x - \lfloor x \rfloor)}{m!} f^{(m)}(x)\,dx$$

Here, B_k are Bernoulli numbers:

- *For* $m = 1$ (trapezoidal rule):

$$\sum_{a \le k < b} f(k) \approx \int_a^b f(x)\,dx - \frac{1}{2}(f(b) - f(a))$$

2.3.3 *Taylor Expansion*

A differentiable function, $G(x)$, can be written as an infinite sum of its derivatives. More specifically, if $G(x)$ is infinitely differentiable and evaluated at a, its Taylor expansion is

$$G(x) = G(a) + \frac{G'(a)}{1!}(x-a) + \frac{G''(a)}{2!}(x-a)^2 + \frac{G'''(a)}{3!}(x-a)^3 + \ldots$$

This expansion is valid within the radius of convergence.

2.3.4 *Law of Large Numbers*

Let $X_1, X_2, \ldots$ be an infinite sequence of independent and identically distributed (i.i.d.) random variables with finite mean μ and variance σ^2. The *law of large numbers (LLN)* states that the sample average:

$$\bar{X}_n = \frac{1}{n}\sum_{i=1}^{n} X_i$$

converges to the expected value μ as $n \to \infty$. This can be expressed as

$$\bar{X}_n \to \mu \quad (\text{as } n \to \infty).$$

2.3.4.1 Variance of the Sample Mean

The variance of the sample mean decreases as the sample size increases:

$$Var(\bar{X}_n) = Var\left(\frac{1}{n}\sum_{i=1}^{n} X_i\right) = \frac{\sigma^2}{n}.$$

$$\begin{aligned} Var(\bar{X}_n) &= Var(\frac{1}{n}(X_1 + \ldots + X_n)) = Var\left(\frac{1}{n}\sum_{i=1}^{n} X_i\right) \\ &= \frac{1}{n^2} Var(X_1 + \ldots + X_n) \\ &= \frac{n\sigma^2}{n^2} = \frac{\sigma^2}{n} \end{aligned}$$

Note The connection between the law of large numbers and the normal distribution lies in the central limit theorem. The CLT states that, regardless of the original distribution of a dataset, the distribution of the sample means will tend to follow a normal distribution as the sample size becomes larger.

The difference between [Weak Law] and [Strong Law] regards the mode of convergence.

2.3.4.2 Weak Law of Large Numbers

The *weak law of large numbers* states that the sample average converges in probability to the expected value:

$$\bar{X}_n \xrightarrow{p} \mu \quad \text{as } n \to \infty$$

Formally, for any $\epsilon > 0$:

$$\lim_{n\to\infty} P(|\bar{X}_n - \mu| > \epsilon) = 0$$

Additionally, the sample mean of an i.i.d. random sample ($\{X_i\}_{i=1}^n$) from any population with a finite mean and variance is a consistent estimator of the population mean μ:

$$plim(\bar{X}_n) = plim\left(\frac{1}{n}\sum_{i=1}^{n} X_i\right) = \mu$$

2.3.4.3 Strong Law of Large Numbers

The *strong law of large numbers* states that the sample average converges almost surely to the expected value:

$$\bar{X}_n \xrightarrow{a.s.} \mu \quad \text{as } n \to \infty.$$

Equivalently, this can be expressed as

$$P\left(\lim_{n\to\infty} \bar{X}_n = \mu\right) = 1.$$

2.3.5 Convergence

2.3.5.1 Convergence in Probability

As $n \to \infty$, an estimator (random variable) θ_n is said to converge in probability to a constant c if:

$$\lim_{n\to\infty} P(|\theta_n - c| \geq \epsilon) = 0 \quad \text{for any } \epsilon > 0.$$

This is denoted as

$$plim(\theta_n) = c \quad \text{or equivalently, } \theta_n \xrightarrow{p} c.$$

Properties of Convergence in Probability

1. **Slutsky's Theorem**: For a continuous function $g(\cdot)$, if $plim(\theta_n) = \theta$, then:

$$plim(g(\theta_n)) = g(\theta).$$

2. If $\gamma_n \xrightarrow{p} \gamma$, then:
   - $plim(\theta_n + \gamma_n) = \theta + \gamma$.
   - $plim(\theta_n \gamma_n) = \theta\gamma$.
   - $plim(\theta_n/\gamma_n) = \theta/\gamma$ (if $\gamma \neq 0$).
3. These properties extend to random vectors and matrices.

2.3.5.2 Convergence in Distribution

As $n \to \infty$, the distribution of a random variable X_n may converge to another ("fixed") distribution. Formally, X_n with CDF $F_n(x)$ converges in distribution to X with CDF $F(x)$ if:

$$\lim_{n\to\infty} |F_n(x) - F(x)| = 0$$

at all points of continuity of $F(x)$. This is denoted as

$$X_n \xrightarrow{d} X \quad \text{or equivalently, } F(x) \text{ is the limiting distribution of } X_n.$$

Asymptotic Properties

- $E(X)$: Limiting mean (asymptotic mean)
- $Var(X)$: Limiting variance (asymptotic variance)

Note Limiting expectations and variances do not necessarily match the expectations and variances of X_n:

$$E(X) \neq \lim_{n\to\infty} E(X_n),$$

$$Avar(X_n) \neq \lim_{n\to\infty} Var(X_n)$$

Properties of Convergence in Distribution

1. **Continuous Mapping Theorem**: For a continuous function $g(\cdot)$, if $X_n \xrightarrow{d} X$, then:

$$g(X_n) \xrightarrow{d} g(X).$$

2. If $Y_n \xrightarrow{d} c$ (a constant), then:
   - $X_n + Y_n \xrightarrow{d} X + c$.
   - $Y_n X_n \xrightarrow{d} cX$.
   - $X_n/Y_n \xrightarrow{d} X/c$ (if $c \neq 0$).
3. These properties also extend to random vectors and matrices (Table 2.7).

Relationship Between Convergence Types

Convergence in probability is stronger than convergence in distribution. Therefore:

- Convergence in distribution does not guarantee convergence in probability.

Table 2.7 Properties of convergence

Convergence in probability	Convergence in distribution
Slutsky's Theorem: For a continuous $g(\cdot)$, if $plim(\theta_n) = \theta$, then $plim(g(\theta_n)) = g(\theta)$	Continuous mapping theorem: For a continuous $g(\cdot)$, if $X_n \xrightarrow{d} X$, then $g(X_n) \xrightarrow{d} g(X)$
If $\gamma_n \xrightarrow{p} \gamma$, then:	If $Y_n \xrightarrow{d} c$, then:
$plim(\theta_n + \gamma_n) = \theta + \gamma$	$X_n + Y_n \xrightarrow{d} X + c$
$plim(\theta_n \gamma_n) = \theta\gamma$	$Y_n X_n \xrightarrow{d} cX$
$plim(\theta_n/\gamma_n) = \theta/\gamma$ (if $\gamma \neq 0$)	$X_n/Y_n \xrightarrow{d} X/c$ (if $c \neq 0$)

2.3.6 *Sufficient Statistics and Likelihood*

2.3.6.1 Likelihood

The *likelihood* describes the degree to which the observed data supports a particular value of a parameter θ:

- The exact value of the likelihood is *not meaningful*; only relative comparisons matter.
- Likelihood is *informative* when comparing parameter values, helping identify which values of θ are more plausible given the data.

For a single observation $Y = y$, the likelihood function is defined as

$$L(\theta_0; y) = P(Y = y \mid \theta = \theta_0) = f_Y(y; \theta_0)$$

where $f_Y(y; \theta_0)$ is the probability density (or mass) function of Y for the parameter θ_0.

Key Insight The likelihood tells us how plausible θ is, given the data we observed. It is *not a probability*, but it is proportional to the probability of observing the data under a given parameter value.

Example Suppose Y follows a binomial distribution with $n = 10$ trials and probability of success p:

$$P(Y = y \mid p) = \binom{10}{y} p^y (1-p)^{10-y}.$$

For $y = 7$ observed successes, the likelihood function becomes

$$L(p; y = 7) = \binom{10}{7} p^7 (1-p)^3$$

We can use this to compare how well different values of p explain the observed data.

2.3.6.2 Likelihood Ratio

The *likelihood ratio* compares the relative likelihood of two parameter values θ_0 and θ_1 given the observed data:

$$\text{Likelihood Ratio} = \frac{L(\theta_0; y)}{L(\theta_1; y)}.$$

- A likelihood ratio greater than 1 implies that θ_0 is more likely than θ_1, given the observed data.
- Likelihood ratios are widely used in hypothesis testing and model comparison to evaluate the evidence against a null hypothesis.

Example For the binomial example above, consider $p_0 = 0.7$ and $p_1 = 0.5$. The likelihood ratio is

$$\frac{L(p_0; y = 7)}{L(p_1; y = 7)} = \frac{\binom{10}{7}(0.7)^7(0.3)^3}{\binom{10}{7}(0.5)^7(0.5)^3}.$$

This simplifies to

$$\frac{(0.7)^7(0.3)^3}{(0.5)^7(0.5)^3}.$$

The likelihood ratio quantifies how much more likely p_0 is compared to p_1 given the observed data.

2.3.6.3 Likelihood Function

For a given sample, the likelihood for all possible values of θ forms the *likelihood function*:

$$L(\theta) = L(\theta; y) = f_Y(y; \theta)$$

For a sample of size n, assuming independence among observations:

$$L(\theta) = \prod_{i=1}^{n} f_Y(y_i; \theta)$$

Taking the natural logarithm of the likelihood gives the *log-likelihood function*:

$$l(\theta) = \sum_{i=1}^{n} \log f_Y(y_i; \theta).$$

Why Log Likelihood?

- The log likelihood simplifies computation by turning products into sums.
- It is particularly useful for optimization, as many numerical methods (e.g., gradient-based algorithms) perform better with sums than products.

Example For $Y_1, Y_2, \ldots, Y_n$ i.i.d. observations from a normal distribution $N(\mu, \sigma^2)$, the likelihood is

$$L(\mu, \sigma^2) = \prod_{i=1}^{n} \frac{1}{\sqrt{2\pi\sigma^2}} \exp\left(-\frac{(y_i - \mu)^2}{2\sigma^2}\right)$$

The log likelihood is

$$l(\mu, \sigma^2) = -\frac{n}{2}\log(2\pi\sigma^2) - \frac{1}{2\sigma^2}\sum_{i=1}^{n}(y_i - \mu)^2$$

2.3.6.4 Sufficient Statistics

A *sufficient statistic* $T(y)$ is a summary of the data that retains all information about a parameter θ. It allows us to focus on this condensed statistic without losing any inferential power regarding θ.

Formal Definition
A statistic $T(y)$ is sufficient for a parameter θ if the conditional probability distribution of the data y, given $T(y)$ and θ, does not depend on θ. Mathematically:

$$P(Y = y \mid T(y), \theta) = P(Y = y \mid T(y)).$$

Alternatively, by the *Factorization Theorem*, $T(y)$ is sufficient if the likelihood can be written as

$$L(\theta; y) = c(y)L^*(\theta; T(y))$$

where:

- $c(y)$ is a function of the data independent of θ.
- $L^*(\theta; T(y))$ is a function that depends on θ and $T(y)$.

In other words, the likelihood function can be rewritten in terms of $T(y)$ alone, without loss of information about θ.

Why Sufficient Statistics Matter
- They allow us to simplify the analysis by reducing the data without losing inferential power.
- Many inferential procedures (e.g., maximum likelihood estimation, Bayesian methods) are simplified by working with sufficient statistics.

Example Consider a sample of i.i.d. observations $Y_1, Y_2, \ldots, Y_n$ from a normal distribution $N(\mu, \sigma^2)$. Here:

1. The sample mean $\bar{Y} = \frac{1}{n}\sum_{i=1}^{n} Y_i$ is sufficient for μ.
2. The sample variance $S^2 = \frac{1}{n-1}\sum_{i=1}^{n}(Y_i - \bar{Y})^2$ is sufficient for σ^2.

Verification The joint density of $y_1, y_2, \ldots, y_n$ can be factored as

$$f(y_1, \ldots, y_n; \mu, \sigma^2) = \underbrace{\frac{1}{(2\pi\sigma^2)^{n/2}} \exp\left(-\frac{1}{2\sigma^2}\sum_{i=1}^{n}(y_i - \bar{y})^2\right)}_{L^*(\mu,\sigma^2;\bar{y},s^2)} \cdot \underbrace{[\text{independent of } \mu,\ \sigma^2]}_{c(y)}.$$

This shows $\bar{Y}$ and S^2 are sufficient.

Usage of Sufficient Statistics

1. **Maximum Likelihood Estimation (MLE):** In MLE, sufficient statistics simplify the optimization problem by reducing the data without losing information.

   Example: In the normal distribution case, μ can be estimated using the sufficient statistic $\bar{Y}$:

$$\hat{\mu}_{MLE} = \bar{Y}.$$

2. **Bayesian Inference:** In Bayesian analysis, the posterior distribution depends on the sufficient statistic rather than the entire dataset. For the normal case:

$$P(\mu \mid \bar{Y}) \propto P(\mu)L(\mu; \bar{Y}).$$

3. **Data Compression:** In practice, sufficient statistics reduce the complexity of data storage and analysis by condensing all relevant information into a smaller representation.

2.3.6.5 Nuisance Parameters

Parameters that are not of direct interest in the analysis but are necessary to model the data are called *nuisance parameters*.

Profile Likelihood To handle nuisance parameters, replace them with their maximum likelihood estimates (MLEs) in the likelihood function, creating a *profile likelihood* for the parameter of interest.

Example of Profile Likelihood
In a regression model with parameters β (coefficients) and σ^2 (error variance), σ^2 is often a nuisance parameter. The profile likelihood for β is obtained by substituting the MLE of σ^2 into the likelihood:

$$L_p(\beta) = L(\beta, \hat{\sigma}^2)$$

where $\hat{\sigma}^2$ is the MLE of σ^2 given β.

This simplifies the problem to focus only on the parameter of interest, β.

2.3.7 *Parameter Transformations*

Transformations of parameters are often used to improve interpretability or statistical properties of models.

2.3.7.1 Log-Odds Transformation

The *log-odds transformation* is commonly used in logistic regression and binary classification problems. It transforms probabilities (which are bounded between 0 and 1) to the real line:

$$\text{Log odds} = g(\theta) = \ln\left(\frac{\theta}{1-\theta}\right)$$

where θ represents a probability (e.g., the success probability in a Bernoulli trial).

2.3.7.2 General Parameter Transformations

For a parameter θ and a transformation $g(\cdot)$:

- If $\theta \in (a, b)$, $g(\theta)$ may map θ to a different range (e.g., $\mathbb{R}$).
- Useful transformations include:
  - Logarithmic: $g(\theta) = \ln(\theta)$ for $\theta > 0$.
  - Exponential: $g(\theta) = e^{\theta}$ for unconstrained θ.
  - Square root: $g(\theta) = \sqrt{\theta}$ for $\theta \geq 0$.

Jacobian Adjustment for Transformations If transforming a parameter in Bayesian inference, the Jacobian of the transformation must be included to ensure proper posterior scaling.

2.3.7.3 Applications of Parameter Transformations

1. **Improving Interpretability**:
   - Probabilities can be transformed to odds or log odds for logistic models.
   - Rates can be transformed logarithmically for multiplicative effects.
2. **Statistical Modeling**:
   - Variance-stabilizing transformations (e.g., log for Poisson data or arcsine for proportions)
   - Regularization or simplification of complex relationships
3. **Optimization**:
   - Transforming constrained parameters (e.g., probabilities or positive scales) to unconstrained scales simplifies optimization algorithms.

2.4 Data Import/Export

R is a powerful and flexible tool for data analysis, but it was originally designed for in-memory statistical computing. This imposes several practical limitations, especially when handling large datasets.

2.4.1 Key Limitations of R

- **Single-Core Default Execution**
  By default, R utilizes only one CPU core, limiting performance for compute-intensive tasks unless parallelization is explicitly implemented.
- **Memory-Based Data Handling**
  R loads data into RAM. This approach becomes problematic when working with datasets that exceed the available memory:
  - **Medium-Sized Files**: Fits within typical RAM (1–2 GB). Processing is straightforward.
  - **Large Files**: Between 2 and 10 GB. May require memory-efficient coding or special packages.
  - **Very Large Files**: Exceed 10 GB. Necessitates distributed or parallel computing solutions.

2.4.2 *Solutions and Workarounds*

1. **Upgrade Hardware**:

- Increase RAM: A simple but often effective solution for moderately large datasets

2. **Leverage High-Performance Computing (HPC) in R**:

There are several HPC strategies and packages in R that facilitate working with large or computationally intensive tasks:

- ***Explicit Parallelism***
  Use packages like `parallel`, `foreach`, `doParallel`, `future`, and `snow` to define how code runs across multiple cores or nodes.
- ***Implicit Parallelism***
  Certain functions in packages such as `data.table`, `dplyr`, or `caret` internally optimize performance across cores when available.
- ***Large-Memory Computation***
  Use memory-efficient structures or external memory algorithms.
- ***MapReduce Paradigm***
  Useful in distributed computing environments, especially with big data infrastructure like Hadoop.

3. **Efficient Data Loading**:

- Limit rows and columns
  Use arguments such as `nrows =` in functions like `read.csv()` or `fread()` to load subsets of large datasets:

4. **Use Specialized Packages for Large Data**:

- In-Memory Matrix Packages (Single Class Type Support)
  These packages interface with C++ to handle large matrices more efficiently:
  - `bigmemory`, `biganalytics`, `bigtabulate`, `synchronicity`, `bigalgebra`, `bigvideo`
- Out-of-Memory Storage (Multiple Class Types)
  For datasets with various data types:
  - `ff` package: Stores data on disk and accesses it as needed, suitable for mixed-type columns.

5. **Handling Very Large Datasets (>10 GB)**

When data size exceeds the capacity of a single machine, distributed computing becomes necessary:

- Hadoop Ecosystem Integration
  - `RHadoop`: A suite of R packages that integrate with the Hadoop framework.
  - `HadoopStreaming`: Enables R scripts to be used as Hadoop mappers and reducers.
  - `Rhipe`: Provides a more R-like interface to Hadoop, using Google's Protocol Buffers for serialization.

2.4.3 Medium Size

```
library("rio")
```

To import multiple files in a directory

```
str(import_list(dir()), which = 1)
```

To export a single data file

```
export(data, "data.csv")
export(data,"data.dta")
export(data,"data.txt")
export(data,"data_cyl.rds")
export(data,"data.rdata")
export(data,"data.R")
export(data,"data.csv.zip")
export(data,"list.json")
```

To export multiple data files

```
export(list(mtcars = mtcars, iris = iris), "data_file_type")
# where data_file_type should substituted with the extension listed above
```

To convert between data file types

```
# convert Stata to SPSS
convert("data.dta", "data.sav")
```

2.4.4 *Large Size*

2.4.4.1 Cloud Computing: Using AWS for Big Data

Amazon Web Service (AWS): Compute resources can be rented at approximately $1/hr. Use AWS to process large datasets without overwhelming your local machine.

2.4.4.2 Importing Large Files as Chunks

```
file_in <- file("in.csv", "r")  # Open a connection to the file
chunk_size <- 100000             # Define chunk size
x <- readLines(file_in, n = chunk_size)  # Read data in chunks
close(file_in)                   # Close the file connection
```

2.4.4.2.1 Using Base R

```
library(data.table)
mydata <- fread("in.csv", header = TRUE)  # Fast and memory-efficient
```

2.4.4.2.2 Using the `data.table` Package

```
library(ff)
x <- read.csv.ffdf(
  file = "file.csv",
  nrow = 10,           # Total rows
  header = TRUE,       # Include headers
  VERBOSE = TRUE,      # Display progress
  first.rows = 10000, # Initial chunk
  next.rows = 50000,  # Subsequent chunks
  colClasses = NA
)
```

2.4.4.2.3 Using the `ff` Package

```
library(bigmemory)
my_data <- read.big.matrix('in.csv', header = TRUE)
```

2.4.4.2.4 Using the bigmemory Package

```
library(sqldf)
my_data <- read.csv.sql('in.csv')

# Example: Filtering during import
iris2 <- read.csv.sql("iris.csv",
    sql = "SELECT * FROM file WHERE Species = 'setosa'")
```

2.4.4.2.5 Using the sqldf Package

```
library(RMySQL)
#> Loading required package: DBI
```

2.4.4.2.6 Using the RMySQL Package

RQLite package

- Download SQLite, pick "A bundle of command-line tools for managing SQLite database files" for Window 10.
- Unzip file, and open `sqlite3.exe`.
- Type in the prompt:
  - `sqlite> .cd 'C:\Users\data'` to specify path to your desired directory
  - `sqlite> .open database_name.db` to open a database
  - To import the CSV file into the database:
    * `sqlite> .mode csv` to specify to SQLite that the next file is .csv file
    * `sqlite> .import file_name.csv datbase_name` to import the csv file to the database
  - `sqlite> .exit` After you are done, exit the sqlite program.

```
library(DBI)
library(dplyr)
library("RSQLite")
setwd("")
con <- dbConnect(RSQLite::SQLite(), "data_base.db")
tbl <- tbl(con, "data_table")
tbl %>%
    filter() %>%
    select() %>%
    collect() # to actually pull the data into the workspace
dbDisconnect(con)
```

2.4.4.2.7 Using the arrow Package

```
library(vroom)

# Import a compressed CSV file
compressed <- vroom_example("mtcars.csv.zip")
data <- vroom(compressed)
```

2.4.4.2.8 Using the vroom Package

```
s = fread("sample.csv")
```

2.4.4.2.9 Using the data.table Package

```
test = ff::read.csv.ffdf(file = "")
object.size(test) # Highest memory usage

test1 = data.table::fread(file = "")
object.size(test1) # Lowest memory usage

test2 = readr::read_csv(file = "")
object.size(test2) # Second lowest memory usage

test3 = vroom::vroom(file = "")
object.size(test3) # Similar to read_csv
```

2.4.4.2.10 Comparisons Regarding Storage Space

When dealing with large datasets—especially those exceeding 10 GB—standard data-loading strategies in R can become impractical due to memory constraints. One common approach is to store these datasets in a compressed format such as `.csv.gz`, which saves disk space while preserving compatibility with many tools.

Compressed files (e.g., `csv.gz`) are especially useful for archiving and transferring large datasets. However, R typically loads the entire dataset into memory before writing or processing, which is inefficient or even infeasible for very large files.

In such cases, *sequential processing* becomes essential. Rather than reading the entire dataset at once, you can process it in chunks or rows, minimizing memory usage (Table 2.8).

While `readr::read_csv()` is faster and more efficient for smaller files, it has a limitation for large files when used with `skip`. The `skip` argument does not

Table 2.8 Comparison between `read.csv()` and `readr::read_csv()`

Function	Supports connections	Sequential access	Performance	Notes
`read.csv()`	Yes	Yes	Slower	Base R; can work with gz-compressed files, and allows looped reading
`readr::read_csv()`	Limited	No	Faster	High performance but rereads lines even when using `skip`

avoid reading the skipped rows—it simply discards them after reading. This leads to redundant I/O operations and significantly slows down performance for large files.

```
readr::read_csv(file, n_max = 100, skip = 0)       # Reads rows 1--100
readr::read_csv(file, n_max = 200, skip = 100)     # Re-reads rows 1--100, then reads 101--300
```

This approach is inefficient when processing data in chunks.

`read.csv()` can read directly from a file or a connection, and unlike `readr::read_csv()`, it *maintains the connection state*, allowing you to loop over chunks without rereading prior rows.

Example: Sequential Reading with Connection

```
con <- gzfile("large_file.csv.gz", open = "rt")  # Text mode
headers <- read.csv(con, nrows = 1)              # Read column names
repeat {
  chunk <- tryCatch(read.csv(con, nrows = 1000), error = function(e) NULL)
  if (is.null(chunk) || nrow(chunk) == 0) break
  # Process 'chunk' here
}
close(con)
```

Occasionally, when reading compressed or encoded files, you might encounter the following error:

Error in (function (con, what, n = 1L, size = NA_integer_, signed = TRUE):
  can only read from a binary connection

This can occur if the file is not correctly interpreted as text. A workaround is to *explicitly set the connection mode to binary* using `"rb"`:

```
con <- gzfile("large_file.csv.gz", open = "rb")  # Binary mode
```

Although `file()` and `gzfile()` generally detect formats automatically, setting the mode explicitly can resolve issues when the behavior is inconsistent.

2.4.4.3 Sequential Processing for Large Data

```
# Open file for sequential reading
file_conn <- file("file.csv", open = "r")
while (TRUE) {
  # Read a chunk of data
  data_chunk <- read.csv(file_conn, nrows = 1000)
  if (nrow(data_chunk) == 0) break  # Stop if no more rows
  # Process the chunk here
}
close(file_conn)  # Close connection
```

2.5 Data Manipulation

```
# Load required packages
library(tidyverse)
#> -- Attaching core tidyverse packages ---- tidyverse 2.0.0 --
#> v dplyr     1.1.4     v readr     2.1.5
#> v forcats   1.0.0     v stringr   1.5.1
#> v ggplot2   3.5.2     v tibble    3.2.1
#> v lubridate 1.9.4     v tidyr     1.3.1
#> v purrr     1.0.4
#> -- Conflicts ---------------------- tidyverse_conflicts() --
#> x dplyr::filter() masks stats::filter()
#> x dplyr::lag()    masks stats::lag()
#> i Use the conflicted package (<http://conflicted.r-lib.org/>) to force all conflicts to become errors
library(lubridate)

# -----------------------------
# Data Structures in R
# -----------------------------

# Create vectors
x <- c(1, 4, 23, 4, 45)
n <- c(1, 3, 5)
g <- c("M", "M", "F")

# Create a data frame
df <- data.frame(n, g)
df  # View the data frame
#>   n g
#> 1 1 M
#> 2 3 M
#> 3 5 F
str(df)  # Check its structure
#> 'data.frame':	3 obs. of  2 variables:
#>  $ n: num  1 3 5
#>  $ g: chr  "M" "M" "F"

# Using tibble for cleaner outputs
df <- tibble(n, g)
df  # View the tibble
#> # A tibble: 3 x 2
#>       n g
#>   <dbl> <chr>
#> 1     1 M
```

```
#> 2      3 M
#> 3      5 F
str(df)
#> tibble [3 x 2] (S3: tbl_df/tbl/data.frame)
#>  $ n: num [1:3] 1 3 5
#>  $ g: chr [1:3] "M" "M" "F"

# Create a list
lst <- list(x, n, g, df)
lst  # Display the list
#> [[1]]
#> [1]  1  4 23  4 45
#>
#> [[2]]
#> [1] 1 3 5
#>
#> [[3]]
#> [1] "M" "M" "F"
#>
#> [[4]]
#> # A tibble: 3 x 2
#>       n g
#>   <dbl> <chr>
#> 1     1 M
#> 2     3 M
#> 3     5 F

# Name list elements
lst2 <- list(num = x, size = n, sex = g, data = df)
lst2  # Named list elements are easier to reference
#> $num
#> [1]  1  4 23  4 45
#>
#> $size
#> [1] 1 3 5
#>
#> $sex
#> [1] "M" "M" "F"
#>
#> $data
#> # A tibble: 3 x 2
#>       n g
#>   <dbl> <chr>
#> 1     1 M
#> 2     3 M
#> 3     5 F

# Another list example with numeric vectors
lst3 <- list(
  x = c(1, 3, 5, 7),
  y = c(2, 2, 2, 4, 5, 5, 5, 6),
  z = c(22, 3, 3, 3, 5, 10)
)
lst3
#> $x
#> [1] 1 3 5 7
#>
#> $y
#> [1] 2 2 2 4 5 5 5 6
#>
#> $z
#> [1] 22  3  3  3  5 10

# Find means of list elements
# One at a time
mean(lst3$x)
```

```
#> [1] 4
mean(lst3$y)
#> [1] 3.875
mean(lst3$z)
#> [1] 7.666667

# Using lapply to calculate means
lapply(lst3, mean)
#> $x
#> [1] 4
#>
#> $y
#> [1] 3.875
#>
#> $z
#> [1] 7.666667

# Simplified output with sapply
sapply(lst3, mean)
#>        x        y        z
#> 4.000000 3.875000 7.666667

# Tidyverse alternative: map() function
map(lst3, mean)
#> $x
#> [1] 4
#>
#> $y
#> [1] 3.875
#>
#> $z
#> [1] 7.666667

# Tidyverse with numeric output: map_dbl()
map_dbl(lst3, mean)
#>        x        y        z
#> 4.000000 3.875000 7.666667

# -----------------------------
# Binding Data Frames
# -----------------------------

# Create tibbles for demonstration
dat01 <- tibble(x = 1:5, y = 5:1)
dat02 <- tibble(x = 10:16, y = x / 2)
dat03 <- tibble(z = runif(5))  # 5 random numbers from (0, 1)

# Row binding
bind_rows(dat01, dat02, dat01)
#> # A tibble: 17 x 2
#>        x     y
#>    <int> <dbl>
#>  1     1   5
#>  2     2   4
#>  3     3   3
#>  4     4   2
#>  5     5   1
#>  6    10   5
#>  7    11   5.5
#>  8    12   6
#>  9    13   6.5
#> 10    14   7
#> 11    15   7.5
#> 12    16   8
#> 13     1   5
#> 14     2   4
```

```
#> 15      3   3
#> 16      4   2
#> 17      5   1

# Add a new identifier column with .id
bind_rows(dat01, dat02, .id = "id")
#> # A tibble: 12 x 3
#>    id        x     y
#>    <chr> <int> <dbl>
#>  1 1         1   5
#>  2 1         2   4
#>  3 1         3   3
#>  4 1         4   2
#>  5 1         5   1
#>  6 2        10   5
#>  7 2        11   5.5
#>  8 2        12   6
#>  9 2        13   6.5
#> 10 2        14   7
#> 11 2        15   7.5
#> 12 2        16   8

# Use named inputs for better identification
bind_rows("dat01" = dat01, "dat02" = dat02, .id = "id")
#> # A tibble: 12 x 3
#>    id        x     y
#>    <chr> <int> <dbl>
#>  1 dat01     1   5
#>  2 dat01     2   4
#>  3 dat01     3   3
#>  4 dat01     4   2
#>  5 dat01     5   1
#>  6 dat02    10   5
#>  7 dat02    11   5.5
#>  8 dat02    12   6
#>  9 dat02    13   6.5
#> 10 dat02    14   7
#> 11 dat02    15   7.5
#> 12 dat02    16   8

# Bind a list of data frames
list01 <- list("dat01" = dat01, "dat02" = dat02)
bind_rows(list01, .id = "source")
#> # A tibble: 12 x 3
#>    source     x     y
#>    <chr>  <int> <dbl>
#>  1 dat01      1   5
#>  2 dat01      2   4
#>  3 dat01      3   3
#>  4 dat01      4   2
#>  5 dat01      5   1
#>  6 dat02     10   5
#>  7 dat02     11   5.5
#>  8 dat02     12   6
#>  9 dat02     13   6.5
#> 10 dat02     14   7
#> 11 dat02     15   7.5
#> 12 dat02     16   8

# Column binding
bind_cols(dat01, dat03)
#> # A tibble: 5 x 3
#>       x     y      z
#>   <int> <int>  <dbl>
#> 1     1     5 0.0495
#> 2     2     4 0.120
```

```r
#> 3     3     3 0.560
#> 4     4     2 0.806
#> 5     5     1 0.709

# ------------------------------
# String Manipulation
# ------------------------------

names <- c("Ford, MS", "Jones, PhD", "Martin, Phd", "Huck, MA, MLS")

# Remove everything after the first comma
str_remove(names, pattern = ", [[:print:]]+")
#> [1] "Ford"   "Jones"  "Martin" "Huck"

# Explanation: [[:print:]]+ matches one or more printable characters

# ------------------------------
# Reshaping Data
# ------------------------------

# Wide format data
wide <- data.frame(
  name = c("Clay", "Garrett", "Addison"),
  test1 = c(78, 93, 90),
  test2 = c(87, 91, 97),
  test3 = c(88, 99, 91)
)

# Long format data
long <- data.frame(
  name = rep(c("Clay", "Garrett", "Addison"), each = 3),
  test = rep(1:3, 3),
  score = c(78, 87, 88, 93, 91, 99, 90, 97, 91)
)

# Summary statistics
aggregate(score ~ name, data = long, mean)  # Mean score per student
#>      name    score
#> 1 Addison 92.66667
#> 2    Clay 84.33333
#> 3 Garrett 94.33333
aggregate(score ~ test, data = long, mean)  # Mean score per test
#>   test    score
#> 1    1 87.00000
#> 2    2 91.66667
#> 3    3 92.66667
```

```r
# Line plot of scores over tests
ggplot(long,
       aes(
           x = factor(test),
           y = score,
           color = name,
           group = name
       )) +
   geom_point() +
   geom_line() +
   xlab("Test") +
   ggtitle("Test Scores by Student")
```

See Fig. 2.22

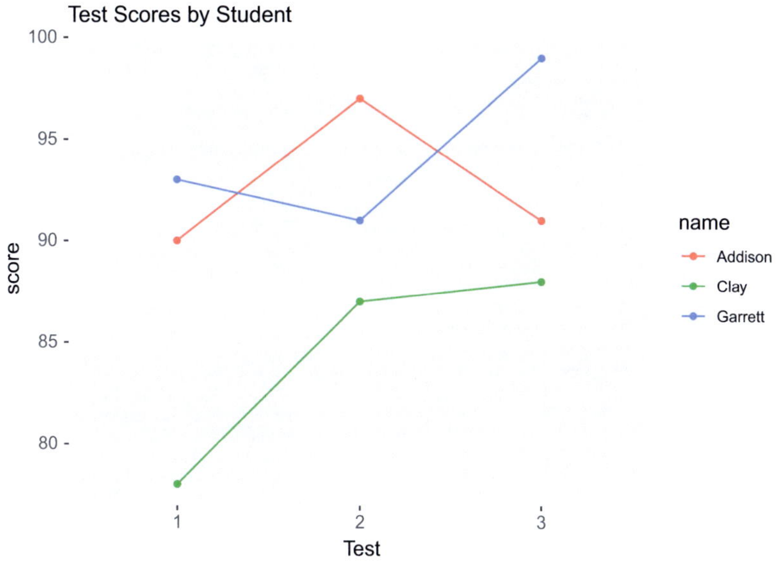

Fig. 2.22 Test scores by student

```
# Reshape wide to long
pivot_longer(wide, test1:test3, names_to = "test", values_to = "score")
#> # A tibble: 9 x 3
#>   name    test  score
#>   <chr>   <chr> <dbl>
#> 1 Clay    test1    78
#> 2 Clay    test2    87
#> 3 Clay    test3    88
#> 4 Garrett test1    93
#> 5 Garrett test2    91
#> 6 Garrett test3    99
#> 7 Addison test1    90
#> 8 Addison test2    97
#> 9 Addison test3    91

# Use names_prefix to clean column names
pivot_longer(
    wide,
    -name,
    names_to = "test",
    values_to = "score",
    names_prefix = "test"
)
#> # A tibble: 9 x 3
#>   name    test  score
#>   <chr>   <chr> <dbl>
#> 1 Clay    1        78
#> 2 Clay    2        87
```

```
#> 3 Clay    3         88
#> 4 Garrett 1         93
#> 5 Garrett 2         91
#> 6 Garrett 3         99
#> 7 Addison 1         90
#> 8 Addison 2         97
#> 9 Addison 3         91

# Reshape long to wide with explicit id_cols argument
pivot_wider(
  long,
  id_cols = name,
  names_from = test,
  values_from = score
)
#> # A tibble: 3 x 4
#>   name      `1`   `2`   `3`
#>   <chr>   <dbl> <dbl> <dbl>
#> 1 Clay       78    87    88
#> 2 Garrett    93    91    99
#> 3 Addison    90    97    91

# Add a prefix to the resulting columns
pivot_wider(
  long,
  id_cols = name,
  names_from = test,
  values_from = score,
  names_prefix = "test"
)
#> # A tibble: 3 x 4
#>   name    test1 test2 test3
#>   <chr>   <dbl> <dbl> <dbl>
#> 1 Clay       78    87    88
#> 2 Garrett    93    91    99
#> 3 Addison    90    97    91
```

The verbs of data manipulation:

- `select`: Selects (or does not select) columns based on their names (e.g., select columns Q1 through Q25).
- `slice`: Selects (or does not select) rows based on their position (e.g., select rows 1:10).
- `mutate`: Adds or derives new columns (or variables) based on existing columns (e.g., creates a new column that expresses measurement in cm based on existing measure in inches).
- `rename`: Renames variables or changes column names (e.g., changes "GraduationRate100" to "grad100").
- `filter`: Selects rows based on a condition (e.g., all rows where gender = Male).
- `arrange`: Orders rows based on variable(s) numeric or alphabetical order (e.g., sort in descending order of Income).

- `sample`: Takes random samples of data (e.g., sample 80% of data to create a "training" set).
- `summarize`: Condenses or aggregates multiple values into single summary values (e.g., calculates median income by age group).
- `group_by`: Converts a tbl into a grouped tbl so that operations are performed "by group"; allows us to summarize data or apply verbs to data by groups (e.g., by gender or treatment).
- The pipe: %>%:
  - Use Ctrl + Shift + M (Win) or Cmd + Shift + M (Mac) to enter in RStudio.
  - The pipe takes the output of a function and "pipes" into the first argument of the next function.
  - New pipe is |>. It should be identical to the old one, except for certain special cases.
- `:=` (Walrus operator): Similar to =, but for cases where you want to use the `glue` package (i.e., dynamic changes in the variable name in the left-hand side).

Writing function in R

Tunneling

`{{` (called curly-curly) allows you to tunnel data variables through arg variables (i.e., function arguments).

```
library(tidyverse)
# ------------------------------
# Writing Functions with {{ }}
# ------------------------------

# Define a custom function using {{ }}
get_mean <- function(data, group_var, var_to_mean) {
  data %>%
    group_by({{group_var}}) %>%
    summarize(mean = mean({{var_to_mean}}, na.rm = TRUE))
}

# Apply the function
data("mtcars")
mtcars %>%
  get_mean(group_var = cyl, var_to_mean = mpg)
#> # A tibble: 3 x 2
#>     cyl  mean
#>   <dbl> <dbl>
#> 1     4  26.7
#> 2     6  19.7
#> 3     8  15.1

# Dynamically name the resulting variable
get_mean <- function(data, group_var, var_to_mean, prefix = "mean_of") {
  data %>%
    group_by({{group_var}}) %>%
    summarize("{prefix}_{{var_to_mean}}" := mean({{var_to_mean}}, na.rm = TRUE))
```

```
}

# Apply the modified function
mtcars %>%
  get_mean(group_var = cyl, var_to_mean = mpg)
#> # A tibble: 3 x 2
#>     cyl mean_of_mpg
#>   <dbl>       <dbl>
#> 1     4        26.7
#> 2     6        19.7
#> 3     8        15.1
```

Chapter 3
Descriptive Statistics

Descriptive statistics serve as the cornerstone of data exploration. This chapter covers the tools used to summarize and visualize data, including measures of central tendency, dispersion, and shape. We emphasize the importance of choosing appropriate graphical techniques to identify patterns, outliers, and relationships. The chapter also includes procedures for assessing normality and a primer on bivariate analysis, laying the groundwork for later inferential methods. Real-world business examples illustrate how descriptive techniques provide insight and guide further analysis (Table 3.1).

When you have an area of interest to research, a problem to solve, or a relationship to investigate, theoretical and empirical processes will help you.

> **Estimand** is defined as "a quantity of scientific interest that can be calculated in the population and does not change its value depending on the data collection design used to measure it (i.e., it does not vary with sample size, survey design, the number of non-respondents, or follow-up efforts)." (Rubin, 1996)

Examples of estimands include:

- Population means
- Population variances
- Correlations
- Factor loadings
- Regression coefficients

3.1 Numerical Measures

Notes

1. Order statistics: $y_{(1)}, y_{(2)}, \ldots, y_{(n)}$, where $y_{(1)} < y_{(2)} < \ldots < y_{(n)}$.

M. Nguyen, *Foundations of Data Analysis*,
https://doi.org/10.1007/978-3-032-01858-8_3

Table 3.1 Differences between a population and a sample

Measures of	Category	Population	Sample
	What is it?	Reality	A small fraction of reality (inference)
	Characteristics described by	Parameters	Statistics
Central tendency	Mean	$\mu = E(Y)$	$\hat{\mu} = \overline{y}$
Central tendency	Median	50th percentile	$y_{(\frac{n+1}{2})}$
Dispersion	Variance	$\sigma^2 = var(Y) = E[(Y-\mu)^2]$	$s^2 = \frac{1}{n-1}\sum_{i=1}^{n}(y_i - \overline{y})^2$
Dispersion	Coefficient of variation	$\frac{\sigma}{\mu}$	$\frac{s}{\overline{y}}$
Dispersion	Interquartile range	Difference between 25th and 75th percentiles; robust to outliers	
Shape	Skewness standardized 3rd central moment (unitless)	$g_1 = \frac{\mu_3}{\sigma^3}$	$\hat{g_1} = \frac{m_3}{m_2^{3/2}}$
Shape	Central moments	$\mu = E(Y), \mu_2 = \sigma^2 = E[(Y-\mu)^2]$, $\mu_3 = E[(Y-\mu)^3]$, $\mu_4 = E[(Y-\mu)^4]$	$m_2 = \frac{1}{n}\sum_{i=1}^{n}(y_i - \overline{y})^2$ $m_3 = \frac{1}{n}\sum_{i=1}^{n}(y_i - \overline{y})^3$
Shape	Kurtosis (peakedness and tail thickness) Standardized 4th central moment	$g_2^* = \frac{E[(Y-\mu)^4]}{\sigma^4}$	$\hat{g_2} = \frac{m_4}{m_2^2} - 3$

2. **Coefficient of Variation**:
   - Defined as the standard deviation divided by the mean.
   - A stable, unitless statistic useful for comparison.
3. **Symmetry**:
   - **Symmetric Distributions**: Mean = Median; Skewness = 0.
   - **Skewed Right**: Mean > Median; Skewness > 0.
   - **Skewed Left**: Mean < Median; Skewness < 0.
4. **Central Moments**:
   - $\mu = E(Y)$
   - $\mu_2 = \sigma^2 = E[(Y-\mu)^2]$
   - $\mu_3 = E[(Y-\mu)^3]$
   - $\mu_4 = E[(Y-\mu)^4]$

Skewness ($\hat{g_1}$)

1. **Sampling Distribution**:
   For samples drawn from a normal population:
   - $\hat{g_1}$ is approximately distributed as $N(0, \frac{6}{n})$ when $n > 150$.
2. **Inference**:
   - **Large Samples**: Inference on skewness can be based on the standard normal distribution. The 95% confidence interval for g_1 is given by:

$$\hat{g_1} \pm 1.96\sqrt{\frac{6}{n}}$$

   - **Small Samples**: For small samples, consult special tables such as:
     - Snedecor and Cochran (1989), Table A 19(i)
     - Monte Carlo test results

Kurtosis ($\hat{g_2}$)

1. **Definitions and Relationships**:
   - A normal distribution has kurtosis $g_2^* = 3$.
     Kurtosis is often redefined as:

$$g_2 = \frac{E[(Y - \mu)^4]}{\sigma^4} - 3$$

where the 4th central moment is estimated by:

$$m_4 = \frac{\sum_{i=1}^{n} (y_i - \overline{y})^4}{n}$$

2. **Sampling Distribution**:
   For large samples ($n > 1000$):
   - $\hat{g_2}$ is approximately distributed as $N(0, \frac{24}{n})$.
3. **Inference**:
   - **Large Samples**: Inference for kurtosis can use standard normal tables (Table 3.2).
   - **Small Samples**: Refer to specialized tables such as:
     - Snedecor and Cochran (1989), Table A 19(ii)
     - Geary (1936)

Table 3.2 Kurtosis types and comparison to the normal distribution

Kurtosis value	Tail behavior	Comparison to normal distribution
$g_2 > 0$ (Leptokurtic)	Heavier tails	Examples: t-distributions
$g_2 < 0$ (Platykurtic)	Lighter tails	Examples: Uniform or certain bounded distributions
$g_2 = 0$ (Mesokurtic)	Normal tails	Exactly matches the normal distribution

```
# Generate random data from a normal distribution
data <- rnorm(100)

# Load the e1071 package for skewness and kurtosis functions
library(e1071)

# Calculate skewness
skewness_value <- skewness(data)
cat("Skewness:", skewness_value, "\n")
#> Skewness: -0.2386494

# Calculate kurtosis
kurtosis_value <- kurtosis(data)
cat("Kurtosis:", kurtosis_value, "\n")
#> Kurtosis: -0.5281164
```

3.2 Graphical Measures

The following table summarizes key graphical measures along with guidance on when and why to use each. More detailed explanations, visual examples, and sample code will be discussed after this table (Table 3.3).

Tips for Selecting the Right Plot:

- **Focus on Your Question**: Are you comparing groups, investigating correlations, or just exploring the overall shape of the data?
- **Match the Plot to Your Data Type**: Continuous vs. categorical data often dictates your choice of chart.
- **Mind the Data Size**: Some plots become cluttered or lose clarity with very large datasets (e.g., stem-and-leaf), while others may be less informative with very few data points.

Table 3.3 Summary of key graphical measures

Graph type	When to use	Why it's useful
Histogram	– Exploring the distribution (shape, center, spread) of a single continuous variable	– Quickly identifies frequency, modes, skewness, and potential outliers – Provides an overview of data "shape"
Box-and-whisker plot	– Comparing the same continuous variable across multiple categories – Identifying median, IQR, and outliers	– Shows distribution at a glance (median, quartiles) – Highlights outliers and potential group differences
Stem-and-leaf plot	– Small, single-variable datasets where you want a textual yet visual distribution view	– Reveals the distribution while preserving actual data values – Easy to spot clusters and gaps for small datasets
Notched boxplot	– Similar to a standard boxplot but with confidence intervals around the median	– If notches don't overlap, it suggests the medians differ significantly – Helps clarify whether differences in medians are likely meaningful
Bagplot (2D boxplot)	– Bivariate data where you want a 2D "boxplot"-style overview – Identifying outliers in two-dimensional space	– Depicts both central region ("bag") and potential outliers – Ideal for discovering clusters or unusual points in two continuous variables
Boxplot matrix	– Multiple continuous variables that you want to compare side-by-side	– Quickly compares distributions of many variables simultaneously – Helpful for spotting differences in median, spread, and outliers
Violin plot	– Same use case as boxplot but you want *more detail* on the distribution's shape	– Combines boxplot features with a density plot – Shows where data are concentrated or sparse within each category

(continued)

Table 3.3 (continued)

Graph type	When to use	Why it's useful
Scatterplot	– Two continuous variables to check for relationships, trends, or outliers	– Visualizes correlation or non-linear patterns – Aids in identifying clusters or extreme values
Pairwise scatterplots	– Initial exploration of *several* variables at once	– Enables a quick scan of relationships between *all* variable pairs – Useful for identifying multivariate patterns or potential correlation structures

3.2.1 Shape

Properly labeling your graphs is essential to ensure that viewers can easily understand the data presented. Below are several examples of graphical measures used to assess the shape of a dataset (Figs. 3.1, 3.2, and 3.3.).

```
# Generate random data for demonstration purposes
data <- rnorm(100)

# Histogram: A graphical representation of the distribution
# of a dataset.
hist(
    data,
    labels = TRUE,
    col = "grey",
    breaks = 12,
    main = "Histogram of Random Data",
    xlab = "Value",
    ylab = "Frequency"
)
```

```
# Interactive Histogram: Using 'highcharter' for
# a more interactive visualization.
# pacman::p_load("highcharter")
# hchart(data, type = "column", name = "Random Data Distribution")
```

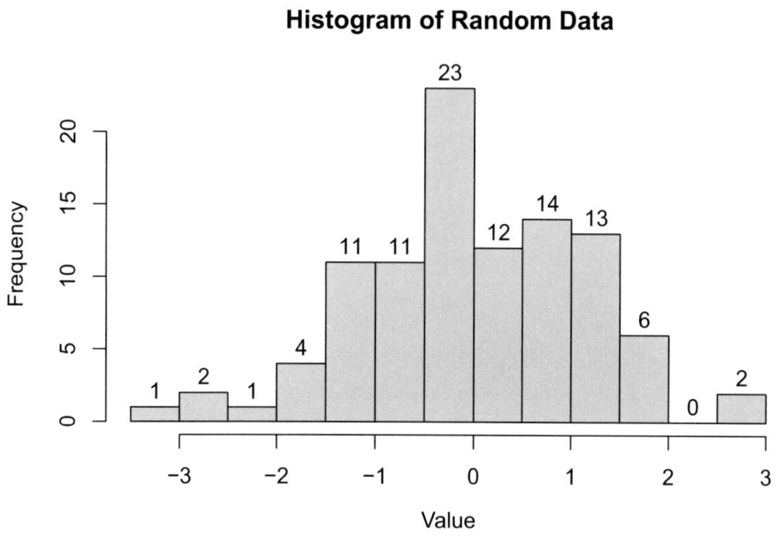

Fig. 3.1 Histogram of random data

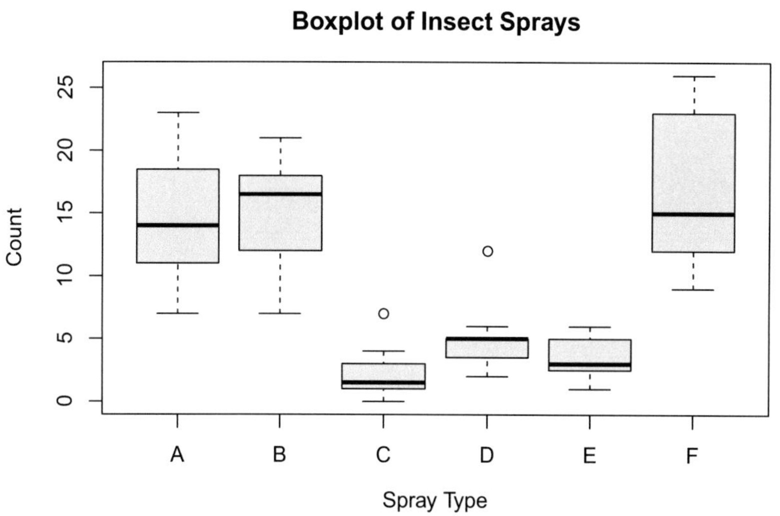

Fig. 3.2 Boxplot of insect sprays

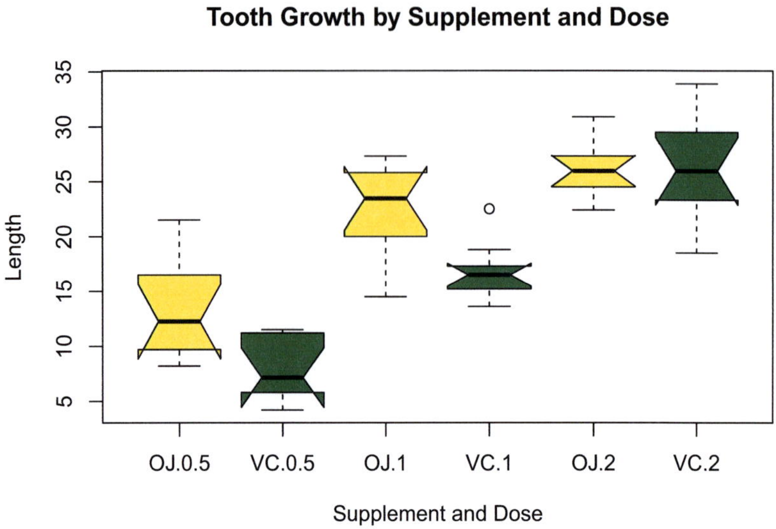

Fig. 3.3 Tooth growth by supplement and dose

```
# Box-and-Whisker Plot: Useful for visualizing the distribution
# and identifying outliers.
boxplot(
    count ~ spray,
    data = InsectSprays,
    col = "lightgray",
    main = "Boxplot of Insect Sprays",
    xlab = "Spray Type",
    ylab = "Count"
)
```

```
# Notched Boxplot: The notches indicate a confidence interval
# around the median.
boxplot(
    len ~ supp * dose,
    data = ToothGrowth,
    notch = TRUE,
    col = c("gold", "darkgreen"),
    main = "Tooth Growth by Supplement and Dose",
    xlab = "Supplement and Dose",
    ylab = "Length"
)
#> Warning in (function (z, notch = FALSE, width = NULL,
#> varwidth = FALSE, : some notches went outside hinges
#> ('box'): maybe set notch=FALSE
```

If the notches of two boxes do not overlap, this suggests that the medians differ significantly (Figs. 3.4, 3.5, 3.6, and 3.7).

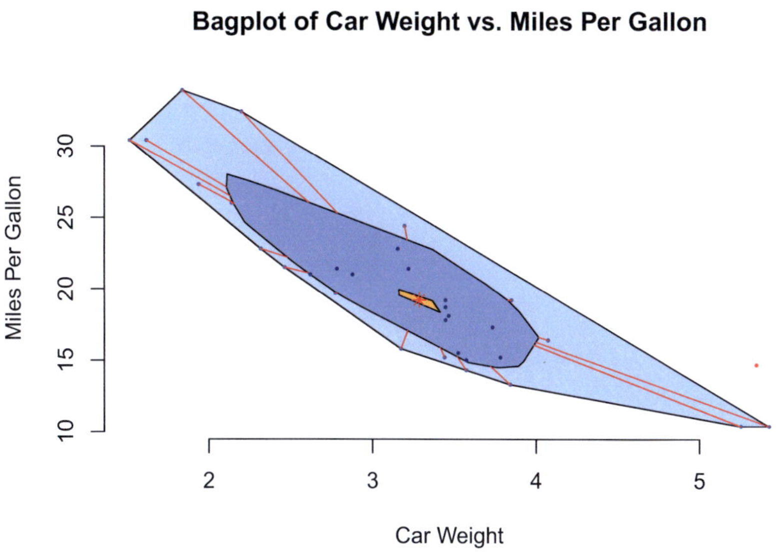

Fig. 3.4 Bagplot of car weight vs. miles per gallon

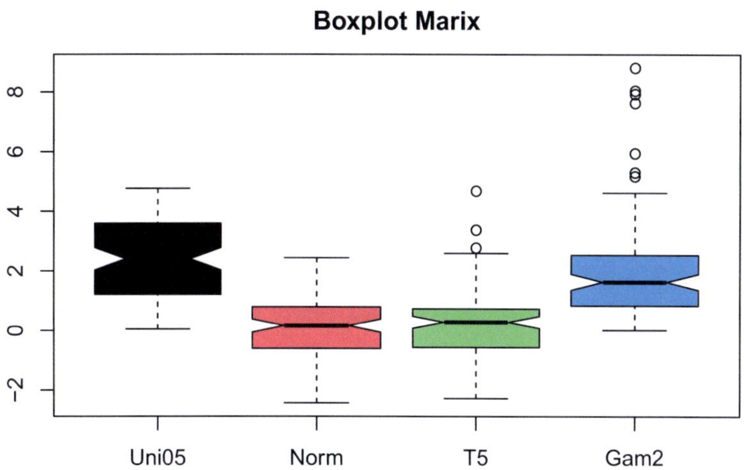

Fig. 3.5 Boxplot marix

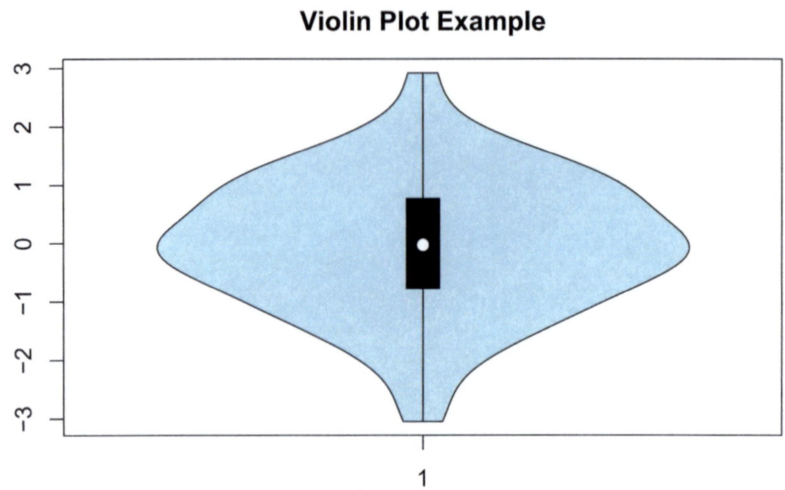

Fig. 3.6 Violin plot example

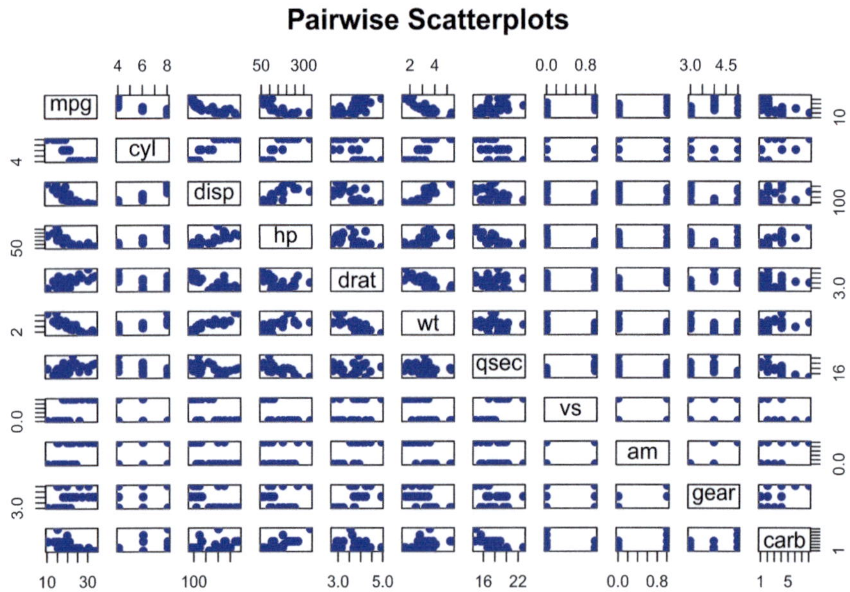

Fig. 3.7 Pairwise scatterplots

```
# Stem-and-Leaf Plot: Provides a quick way to visualize
# the distribution of data.
stem(data)
#>
#>   The decimal point is at the |
#>
#>   -3 | 1
#>   -2 | 8710
#>   -1 | 986555432221100
#>   -0 | 998888876655444432222211110000000
#>    0 | 0000111233456667777788899
#>    1 | 111111222244457777
#>    2 | 079
```

```
# Bagplot - A 2D Boxplot Extension: Visualizes the spread and
# identifies outliers in two-dimensional data.
pacman::p_load(aplpack)
attach(mtcars)
bagplot(wt,
        mpg,
        xlab = "Car Weight",
        ylab = "Miles Per Gallon",
        main = "Bagplot of Car Weight vs. Miles Per Gallon")
```

```
detach(mtcars)
```

Below are some advanced plot types that can provide deeper insights into data:

```
# boxplot.matrix(): Creates boxplots for each column in a matrix.
# Useful for comparing multiple variables.
graphics::boxplot.matrix(
    cbind(
        Uni05 = (1:100) / 21,
        Norm = rnorm(100),
        T5 = rt(100, df = 5),
        Gam2 = rgamma(100, shape = 2)
    ),
    main = "Boxplot Marix",
    notch = TRUE,
    col = 1:4
)
```

```
# Violin Plot (vioplot()): Combines a boxplot with a density plot,
# providing more information about the distribution.
library("vioplot")
#> Loading required package: sm
#> Package 'sm', version 2.2-6.0: type help(sm) for summary information
#> Loading required package: zoo
#>
#> Attaching package: 'zoo'
```

```
#> The following objects are masked from 'package:base':
#>
#>     as.Date, as.Date.numeric
vioplot(data, col = "lightblue", main = "Violin Plot Example")
```

3.2.2 *Scatterplot*

Scatterplots are useful for visualizing relationships between two continuous variables. They help identify patterns, correlations, and outliers.

Pairwise Scatterplots: Visualizes relationships between all pairs of variables in a dataset. This is especially useful for exploring potential correlations.

```
pairs(mtcars,
      main = "Pairwise Scatterplots",
      pch = 19,
      col = "blue")
```

3.3 Normality Assessment

The Normal (Gaussian) distribution plays a critical role in statistical analyses due to its theoretical and practical applications. Many statistical methods assume normality in the data, making it essential to assess whether our variable of interest follows a normal distribution. To achieve this, we utilize both Numerical Measures and Graphical Assessment (Fig. 3.8).

3.3.1 *Graphical Assessment*

Graphical methods provide an intuitive way to visually inspect the normality of a dataset. One of the most common methods is the **Q-Q plot** (quantile-quantile plot). The Q-Q plot compares the quantiles of the sample data to the quantiles of a theoretical normal distribution. Deviations from the line indicate departures from normality.

Below is an example of using the `qqnorm` and `qqline` functions in R to assess the normality of the `precip` dataset, which contains precipitation data (in inches per year) for 70 U.S. cities:

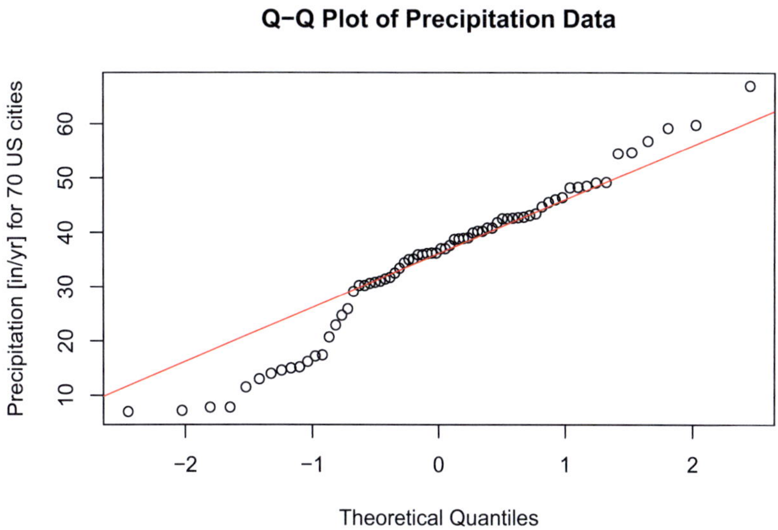

Fig. 3.8 Q-Q plot of precipitation data

```r
# Load the required package
pacman::p_load("car")

# Generate a Q-Q plot
qqnorm(precip,
       ylab = "Precipitation [in/yr] for 70 US cities",
       main = "Q-Q Plot of Precipitation Data")
qqline(precip, col = "red")
```

Interpretation

- **Theoretical Line**: The red line represents the expected relationship if the data were perfectly normally distributed.
- **Data Points**: The dots represent the actual empirical data.

If the points closely align with the theoretical line, we can conclude that the data likely follow a normal distribution. However, noticeable deviations from the line, particularly systematic patterns (e.g., curves or s-shaped patterns), indicate potential departures from normality.

Tips

1. **Small Deviations**: Minor deviations from the line in small datasets are not uncommon and may not significantly impact analyses that assume normality.
2. **Systematic Patterns**: Look for clear trends, such as clusters or s-shaped curves, which suggest skewness or heavy tails.
3. **Complementary Tests**: Always pair graphical methods with numerical measures (e.g., Shapiro-Wilk test) to make a robust conclusion.

When interpreting a Q-Q plot, it is helpful to see both ideal and non-ideal scenarios. Below is an illustrative example:

1. **Normal Data**: Points fall closely along the line.
2. **Skewed Data**: Points systematically deviate from the line, curving upward or downward.
3. **Heavy Tails**: Points deviate at the extremes (ends) of the distribution.

By combining visual inspection and numerical measures, we can better understand the nature of our data and its alignment with the assumption of normality.

3.3.2 Summary Statistics

While graphical assessments, such as Q-Q plots, provide a visual indication of normality, they may not always offer a definitive conclusion. To supplement graphical methods, statistical tests are often employed. These tests provide quantitative evidence to support or refute the assumption of normality. The most common methods can be classified into two categories:

- Methods Based on Normal Probability Plot
  - Correlation Coefficient with Normal Probability Plots
  - Shapiro-Wilk Test
- Methods based on empirical cumulative distribution function
  - Anderson-Darling Test
  - Kolmogorov-Smirnov Test
  - Cramer-von Mises Test
  - Jarque–Bera Test

3.3.2.1 Methods Based on Normal Probability Plot

3.3.2.1.1 Correlation Coefficient with Normal Probability Plots

As described by Looney and Gulledge Jr (1985) and Shapiro and Francia (1972), this method evaluates the linearity of a normal probability plot by calculating the correlation coefficient between the ordered sample values $y_{(i)}$ and their theoretical normal quantiles m_i^*. A perfectly linear relationship suggests that the data follow a normal distribution.

The correlation coefficient, denoted W^*, is given by:

$$W^* = \frac{\sum_{i=1}^{n}(y_{(i)} - \bar{y})(m_i^* - 0)}{\sqrt{\sum_{i=1}^{n}(y_{(i)} - \bar{y})^2 \cdot \sum_{i=1}^{n}(m_i^* - 0)^2}}$$

where:

- $\bar{y}$ is the sample mean,
- $\bar{m}^* = 0$ under the null hypothesis of normality.

The **Pearson product-moment correlation formula** can also be used to evaluate this relationship:

$$\hat{\rho} = \frac{\sum_{i=1}^{n}(y_i - \bar{y})(x_i - \bar{x})}{\sqrt{\sum_{i=1}^{n}(y_i - \bar{y})^2 \cdot \sum_{i=1}^{n}(x_i - \bar{x})^2}}$$

- **Interpretation**:
  - When the correlation is 1, the plot is exactly linear, and normality is assumed.
  - The closer the correlation is to 0, the stronger the evidence to reject normality.
  - Inference on W^* requires reference to special tables (Looney and Gulledge Jr, 1985).

```
library("EnvStats")
#>
#> Attaching package: 'EnvStats'
#> The following object is masked from 'package:car':
#>
#>     qqPlot
#> The following objects are masked from 'package:e1071':
#>
#>     kurtosis, skewness
#> The following objects are masked from 'package:stats':
#>
#>     predict, predict.lm
#> The following object is masked from 'package:base':
#>
#>     print.default

# Perform Probability Plot Correlation Coefficient (PPCC) Test
gofTest(data, test = "ppcc")$p.value # Probability Plot Correlation Coefficient
#> [1] 0.6860369
```

3.3.2.1.2 Shapiro-Wilk Test

The Shapiro-Wilk test (Shapiro and Wilk, 1965) is one of the most widely used tests for assessing normality, especially for sample sizes $n < 2000$. This test evaluates how well the data's order statistics match a theoretical normal distribution. The test statistic, W, is computed as:

$$W = \frac{\sum_{i=1}^{n} a_i x_{(i)}}{\sum_{i=1}^{n}(x_{(i)} - \bar{x})^2}$$

where

- n: The sample size.
- $x_{(i)}$: The i-th smallest value in the sample (the ordered data).
- $\bar{x}$: The sample mean.
- a_i: Weights derived from the expected values and variances of the order statistics of a normal distribution, precomputed based on the sample size n.

Sensitive to:

1. **Symmetry**
   - The Shapiro-Wilk test assesses whether a sample is drawn from a normal distribution, which assumes **symmetry** around the mean.
   - If the data exhibit skewness (a lack of symmetry), the test is likely to reject the null hypothesis of normality.
2. **Heavy Tails**
   - Heavy tails refer to distributions where extreme values (outliers) are more likely compared to a normal distribution.
   - The Shapiro-Wilk test is also sensitive to such departures from normality because heavy tails affect the spread and variance, which are central to the calculation of the test statistic W.

Hence, the Shapiro-Wilk test's sensitivity to these deviations makes it a powerful tool for detecting non-normality only in small to moderate-sized samples. However:

- It is generally more sensitive to **symmetry** (skewness) than to **tail behavior** (kurtosis).
- In very large samples, even small deviations in symmetry or tail behavior may cause the test to reject the null hypothesis, even if the data is practically "normal" for the intended analysis.
  - Small sample sizes may lack power to detect deviations from normality.
  - Large sample sizes may detect minor deviations that are not practically significant.

Key Steps:

1. **Sort the Data:** Arrange the sample data in ascending order, yielding $x_{(1)}, x_{(2)}, \ldots, x_{(n)}$.
2. **Compute Weights:** The weights a_i are determined using a covariance matrix of the normal order statistics. These are optimized to maximize the power of the test.
3. **Calculate** W: Use the formula to determine W, which ranges from 0 to 1.

Decision Rule:

- **Null Hypothesis** (H_0): The data follows a normal distribution.
- **Alternative Hypothesis** (H_1): The data does not follow a normal distribution.
- A small W value, along with a p-value below a chosen significance level (e.g., 0.05), leads to rejection of H_0.
  - Under normality, W approaches 1.
  - Smaller values of W indicate deviations from normality.

```
# Perform Shapiro-Wilk Test (Default for gofTest)
EnvStats::gofTest(mtcars$mpg, test = "sw")
#> $distribution
#> [1] "Normal"
#>
#> $dist.abb
#> [1] "norm"
#>
#> $distribution.parameters
#>      mean        sd
#> 20.090625  6.026948
#>
#> $n.param.est
#> [1] 2
#>
#> $estimation.method
#> [1] "mvue"
#>
#> $statistic
#>         W
#> 0.9475647
#>
#> $sample.size
#> [1] 32
#>
#> $parameters
#>  n
#> 32
#>
#> $z.value
#> [1] 1.160703
#>
#> $p.value
#> [1] 0.1228814
#>
#> $alternative
#> [1] "True cdf does not equal the\n                Normal Distribution."
#>
#> $method
#> [1] "Shapiro-Wilk GOF"
#>
#> $data
```

```
#>  [1] 21.0 21.0 22.8 21.4 18.7 18.1 14.3 24.4 22.8 19.2 17.8
#> [12] 16.4 17.3 15.2 10.4 10.4 14.7 32.4 30.4 33.9 21.5 15.5
#> [23] 15.2 13.3 19.2 27.3 26.0 30.4 15.8 19.7 15.0 21.4
#>
#> $data.name
#> [1] "mtcars$mpg"
#>
#> $bad.obs
#> [1] 0
#>
#> attr(,"class")
#> [1] "gof"
```

3.3.2.2 Methods Based on Empirical Cumulative Distribution Function

The **Empirical Cumulative Distribution Function (ECDF)** is a way to represent the distribution of a sample dataset in cumulative terms. It answers the question:

> "What fraction of the observations in my dataset are less than or equal to a given value x?"

The ECDF is defined as:

$$F_n(x) = \frac{1}{n} \sum_{i=1}^{n} \mathbb{I}(X_i \leq x)$$

where: ;

- $F_(x)$: ECDF at value x.
- n: Total number of data points.
- $\mathbb{I}(X_i \leq x)$: Indicator function, equal to 1 if $X_i \leq x$, otherwise 0.

This method is especially useful for large sample sizes and can be applied to distributions beyond the normal (Gaussian) distribution (Fig. 3.9).

Properties of the ECDF

1. **Step Function**: The ECDF is a step function that increases by $1/n$ at each data point.
2. **Non-decreasing**: As x increases, $F_n(x)$ never decreases.
3. **Range**: The ECDF starts at 0 and ends at 1:
   - $F_n(x) = 0$ for $x < \min(X)$.
   - $F_n(x) = 1$ for $x \geq \max(X)$.
4. **Convergence**: As $n \to \infty$, the ECDF approaches the true cumulative distribution function (CDF) of the population (Fig. 3.10).

Let's consider a sample dataset $\{3, 7, 7, 10, 15\}$. The ECDF at different values of x is calculated as (Table 3.4):

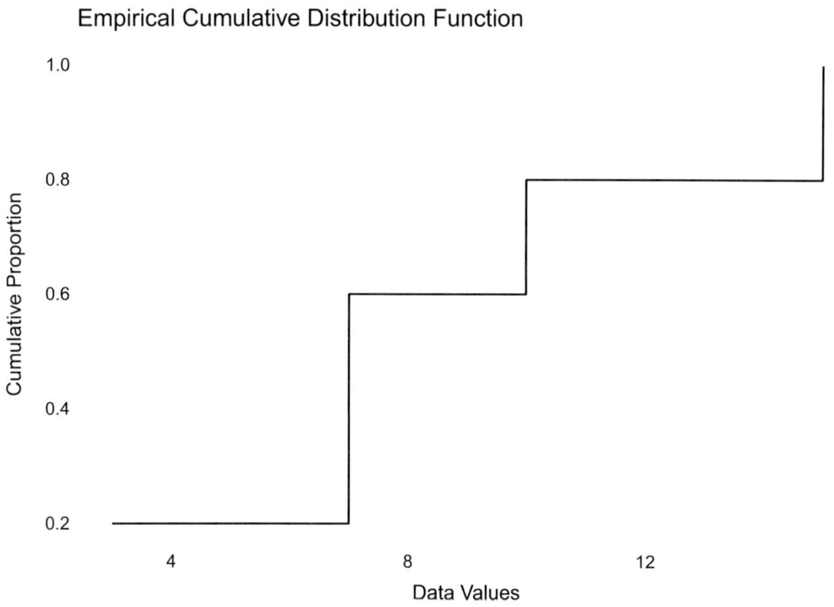

Fig. 3.9 Empirical cumulative distribution function

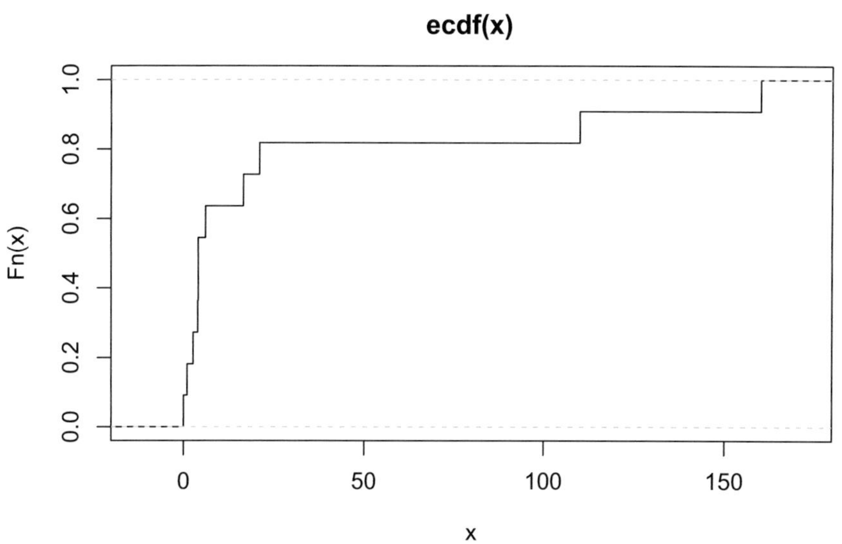

Fig. 3.10 Empirical cumulative distribution function alternative

Table 3.4 Empirical cumulative distribution function computation example

x	$\mathbb{I}(X_i \leq x)$ for each X_i	Count $\leq x$	ECDF $F_n(x)$
$x = 5$	$\{1, 0, 0, 0, 0\}$	1	$1/5 = 0.2$
$x = 7$	$\{1, 1, 1, 0, 0\}$	3	$3/5 = 0.6$
$x = 12$	$\{1, 1, 1, 1, 0\}$	4	$4/5 = 0.8$
$x = 15$	$\{1, 1, 1, 1, 1\}$	5	$5/5 = 1.0$

Applications of the ECDF

1. **Goodness-of-fit Tests**: Compare the ECDF to a theoretical CDF (e.g., using the Kolmogorov-Smirnov test).
2. **Outlier Detection**: Analyze cumulative trends to spot unusual data points.
3. **Visual Data Exploration**: Use the ECDF to understand the spread, skewness, and distribution of the data.
4. **Comparing Distributions**: Compare the ECDFs of two datasets to assess differences in their distributions.

```r
# Load required libraries
library(ggplot2)

# Sample dataset
data <- c(3, 7, 7, 10, 15)

# ECDF calculation
ecdf_function <- ecdf(data)

# Generate a data frame for plotting
ecdf_data <- data.frame(x = sort(unique(data)),
                        ecdf = sapply(sort(unique(data)), function(x)
                          mean(data <= x)))

# Display ECDF values
print(ecdf_data)
#>    x ecdf
#> 1  3  0.2
#> 2  7  0.6
#> 3 10  0.8
#> 4 15  1.0
```

```r
# Plot the ECDF
ggplot(ecdf_data, aes(x = x, y = ecdf)) +
  geom_step() +
  labs(
    title = "Empirical Cumulative Distribution Function",
    x = "Data Values",
    y = "Cumulative Proportion"
  ) +
  theme_minimal()
```

```
# Alternatively
plot.ecdf(as.numeric(mtcars[1, ]),
          verticals = TRUE,
          do.points = FALSE)
```

3.3.2.2.1 Anderson-Darling Test

The **Anderson-Darling test** statistic (Anderson and Darling, 1952) is given by:

$$A^2 = \int_{-\infty}^{\infty} \frac{(F_n(t) - F(t))^2}{F(t)(1 - F(t))} dF(t)$$

This test calculates a weighted average of squared deviations between the empirical cumulative distribution function (CDF), $F_n(t)$, and the theoretical CDF, $F(t)$. More weight is given to deviations in the tails of the distribution, which makes the test particularly sensitive to these regions.

For a sample of size n, with ordered observations $y_{(1)}, y_{(2)}, \ldots, y_{(n)}$, the Anderson-Darling test statistic can also be written as:

$$A^2 = -n - \frac{1}{n} \sum_{i=1}^{n} \left[(2i - 1) \ln(F(y_{(i)})) + (2n + 1 - 2i) \ln(1 - F(y_{(i)}))\right]$$

For the **normal distribution**, the test statistic is further simplified. Using the transformation:

$$p_i = \Phi\left(\frac{y_{(i)} - \bar{y}}{s}\right),$$

where:

- p_i is the cumulative probability under the standard normal distribution,
- $y_{(i)}$ are the ordered sample values,
- $\bar{y}$ is the sample mean,
- s is the sample standard deviation,

the formula becomes:

$$A^2 = -n - \frac{1}{n} \sum_{i=1}^{n} [(2i - 1) \ln(p_i) + (2n + 1 - 2i) \ln(1 - p_i)].$$

Key Features of the Test

1. **CDF-Based Weighting**: The Anderson-Darling test gives more weight to deviations in the tails, which makes it particularly sensitive to detecting non-normality in these regions.
2. **Sensitivity**: Compared to other goodness-of-fit tests, such as the Kolmogorov-Smirnov Test, the Anderson-Darling test is better at identifying differences in the tails of the distribution.
3. **Integral Form**: The test statistic can also be expressed as an integral over the theoretical CDF:

$$A^2 = n \int_{-\infty}^{\infty} \frac{[F_n(t) - F(t)]^2}{F(t)(1 - F(t))} dF(t),$$

where $F_n(t)$ is the empirical CDF, and $F(t)$ is the specified theoretical CDF.
4. **Applications**:
   - Testing for normality or other distributions (e.g., exponential, Weibull).
   - Validating assumptions in statistical models.
   - Comparing data to theoretical distributions.

Hypothesis Testing

- **Null Hypothesis** (H_0): The data follows the specified distribution (e.g., normal distribution).
- **Alternative Hypothesis** (H_1): The data does not follow the specified distribution.
- The null hypothesis is rejected if A^2 is too large, indicating a poor fit to the specified distribution.

Critical values for the test statistic are provided by (Marsaglia and Marsaglia, 2004) and (Stephens, 1974).

Applications to Other Distributions

The Anderson-Darling test can be applied to various distributions by using specific transformation methods. Examples include:

- **Exponential**
- **Logistic**
- **Gumbel**
- **Extreme-value**
- **Weibull** (after logarithmic transformation: $\log(\text{Weibull}) = \text{Gumbel}$)
- **Gamma**
- **Cauchy**
- **von Mises**
- **Log-normal (two-parameter)**

For more details on transformations and critical values, consult (Stephens, 1974).

```
# Perform Anderson-Darling Test
library(nortest)
ad_test_result <- ad.test(mtcars$mpg)

# Output the test statistic and p-value
ad_test_result
#>
#>  Anderson-Darling normality test
#>
#> data:  mtcars$mpg
#> A = 0.57968, p-value = 0.1207
```

Alternatively, for a broader range of distributions, use the `gofTest` function from the `gof` package:

```
# General goodness-of-fit test with Anderson-Darling
library(EnvStats)
gof_test_result <- EnvStats::gofTest(mtcars$mpg, test = "ad")

# Extract the p-value
gof_test_result$p.value
#> [1] 0.1207371
```

3.3.2.2.2 Kolmogorov-Smirnov Test

The **Kolmogorov-Smirnov (K-S) test** is a nonparametric test that compares the empirical cumulative distribution function (ECDF) of a sample to a theoretical cumulative distribution function (CDF), or compares the ECDFs of two samples. It is used to assess whether a sample comes from a specific distribution (one-sample test) or to compare two samples (two-sample test).

The test statistic D_n for the one-sample test is defined as:

$$D_n = \sup_x |F_n(x) - F(x)|,$$

where:

- $F_n(x)$ is the empirical CDF of the sample,
- $F(x)$ is the theoretical CDF under the null hypothesis,
- $\sup_x$ denotes the supremum (largest value) over all possible values of x.

For the two-sample K-S test, the statistic is:

$$D_{n,m} = \sup_x \left| F_{n,1}(x) - F_{m,2}(x) \right|,$$

where $F_{n,1}(x)$ and $F_{m,2}(x)$ are the empirical CDFs of the two samples, with sizes n and m, respectively.

Hypotheses

- **Null hypothesis** (H_0): The sample comes from the specified distribution (one-sample) or the two samples are drawn from the same distribution (two-sample).
- **Alternative hypothesis** (H_1): The sample does not come from the specified distribution (one-sample) or the two samples are drawn from different distributions (two-sample).

Properties

1. **Based on the Largest Deviation**: The K-S test is sensitive to the largest absolute difference between the empirical and expected CDFs, making it effective for detecting shifts in location or scale.
2. **Distribution-Free**: The test does not assume a specific distribution for the data under the null hypothesis. Its significance level is determined from the distribution of the test statistic under the null hypothesis.
3. **Limitations**:
   - The test is more sensitive near the center of the distribution than in the tails.
   - It may not perform well with discrete data or small sample sizes.
4. **Related Tests**:
   - **Kuiper's Test**: A variation of the K-S test that is sensitive to deviations in both the center and tails of the distribution. The Kuiper test statistic is:

$$V_n = D^+ + D^-,$$

where D^+ and D^- are the maximum positive and negative deviations of the empirical CDF from the theoretical CDF.

Applications

- Testing for normality or other specified distributions.
- Comparing two datasets to determine if they are drawn from the same distribution.

To perform a one-sample K-S test in R, use the `ks.test()` function. To check the goodness of fit for a specific distribution, the `gofTest()` function from a package like `DescTools` can also be used.

```
# One-sample Kolmogorov-Smirnov test for normality
data <- rnorm(50)  # Generate random normal data
ks.test(data, "pnorm", mean(data), sd(data))
#>
#>  Exact one-sample Kolmogorov-Smirnov test
#>
#> data:  data
#> D = 0.12483, p-value = 0.3852
#> alternative hypothesis: two-sided
```

```
# Goodness-of-fit test using gofTest
library(DescTools)
#> Registered S3 method overwritten by 'DescTools':
#>   method         from
#>   plot.bagplot aplpack
#>
#> Attaching package: 'DescTools'
#> The following object is masked from 'package:car':
#>
#>     Recode
#> The following object is masked from 'package:aplpack':
#>
#>     plot.bagplot
gofTest(data, test = "ks")$p.value  # Kolmogorov-Smirnov test p-value
#> Warning in ksGofTest(x = c(0.659408691766304, 1.85651542439068,
#> -0.217309858510989, : The standard Kolmogorov-Smirnov test is very
#> conservative (Type I error smaller than assumed; high Type II error)
#> for testing departures from the Normal distribution when you have to
#> estimate the distribution parameters.
#> [1] 0.3852307
```

- **Advantages**:
  - Simple and widely applicable.
  - Distribution-free under the null hypothesis.
- **Limitations**:
  - Sensitive to sample size: small deviations may lead to significance in large samples.
  - Reduced sensitivity to differences in the tails compared to the Anderson-Darling test.

The Kolmogorov-Smirnov test provides a general-purpose method for goodness-of-fit testing and sample comparison, with particular utility in detecting central deviations.

3.3.2.2.3 Cramer-von Mises Test

The **Cramer-von Mises (CVM) test** is a nonparametric goodness-of-fit test that evaluates the agreement between the empirical cumulative distribution function (ECDF) of a sample and a specified theoretical cumulative distribution function (CDF). Unlike the Kolmogorov-Smirnov test, which focuses on the largest discrepancy, the Cramer-von Mises test considers the **average squared discrepancy** across the entire distribution. Unlike the Anderson-Darling test, it weights all parts of the distribution equally.

The test statistic W^2 for the one-sample Cramer-von Mises test is defined as:

$$W^2 = n \int_{-\infty}^{\infty} [F_n(t) - F(t)]^2 \, dF(t),$$

where:

- $F_n(t)$ is the empirical CDF,
- $F(t)$ is the specified theoretical CDF under the null hypothesis,
- n is the sample size.

In practice, W^2 is computed using the ordered sample values $y_{(1)}, y_{(2)}, \ldots, y_{(n)}$ as:

$$W^2 = \sum_{i=1}^{n} \left(F(y_{(i)}) - \frac{2i-1}{2n} \right)^2 + \frac{1}{12n},$$

where:

- $F(y_{(i)})$ is the theoretical CDF evaluated at the ordered sample values $y_{(i)}$.

Hypotheses

- **Null hypothesis** (H_0): The sample data follow the specified distribution.
- **Alternative hypothesis** (H_1): The sample data do not follow the specified distribution.

Properties

1. **Focus on Average Discrepancy**: The Cramer-von Mises test measures the overall goodness-of-fit by considering the squared deviations across all points in the distribution, ensuring equal weighting of discrepancies.
2. **Comparison to Anderson-Darling**: Unlike the Anderson-Darling test, which gives more weight to deviations in the tails, the CVM test weights all parts of the distribution equally.
3. **Integral Representation**: The statistic is expressed as an integral over the squared differences between the empirical and theoretical CDFs.
4. **Two-Sample Test**: The Cramer-von Mises framework can also be extended to compare two empirical CDFs. The two-sample statistic is based on the pooled empirical CDF.

Applications

- Assessing goodness-of-fit for a theoretical distribution (e.g., normal, exponential, Weibull).
- Comparing two datasets to determine if they are drawn from similar distributions.
- Validating model assumptions.

To perform a Cramer-von Mises test in R, the `gofTest()` function from the `DescTools` package can be used. Below is an example:

```
# Generate random normal data
data <- rnorm(50)

# Perform the Cramer-von Mises test
library(DescTools)
gofTest(data, test = "cvm")$p.value  # Cramer-von Mises test p-value
#> [1] 0.3283611
```

- **Advantages**:
  - Considers discrepancies across the entire distribution.
  - Robust to outliers due to equal weighting.
  - Simple to compute and interpret.
- **Limitations**:
  - Less sensitive to deviations in the tails compared to the Anderson-Darling test.
  - May be less powerful than the Kolmogorov-Smirnov test in detecting central shifts.

3.3.2.2.4 Jarque-Bera Test

The **Jarque-Bera (JB) test** (Bera and Jarque, 1981) is a goodness-of-fit test used to check whether a dataset follows a normal distribution. It is based on the **skewness** and **kurtosis** of the data, which measure the asymmetry and the "tailedness" of the distribution, respectively.

The Jarque-Bera test statistic is defined as:

$$JB = \frac{n}{6}\left(S^2 + \frac{(K-3)^2}{4}\right),$$

where:

- n is the sample size,
- S is the sample skewness,
- K is the sample kurtosis.

Skewness (S) is calculated as:

$$S = \frac{\hat{\mu}_3}{\hat{\sigma}^3} = \frac{\frac{1}{n}\sum_{i=1}^{n}(x_i - \bar{x})^3}{\left(\frac{1}{n}\sum_{i=1}^{n}(x_i - \bar{x})^2\right)^{3/2}},$$

where:

- $\hat{\mu}_3$ is the third central moment,
- $\hat{\sigma}$ is the standard deviation,
- $\bar{x}$ is the sample mean.

Kurtosis (K) is calculated as:

$$K = \frac{\hat{\mu}_4}{\hat{\sigma}^4} = \frac{\frac{1}{n}\sum_{i=1}^{n}(x_i - \bar{x})^4}{\left(\frac{1}{n}\sum_{i=1}^{n}(x_i - \bar{x})^2\right)^2},$$

where:

- $\hat{\mu}_4$ is the fourth central moment.

Hypothesis

- **Null hypothesis** (H_0): The data follow a normal distribution, implying:
  - Skewness $S = 0$,
  - Excess kurtosis $K - 3 = 0$.
- **Alternative hypothesis** (H_1): The data do not follow a normal distribution.

Distribution of the JB Statistic
Under the null hypothesis, the Jarque-Bera statistic asymptotically follows a chi-squared distribution with 2 degrees of freedom:

$$JB \sim \chi_2^2.$$

Properties

1. **Sensitivity**:
   - Skewness (S) captures asymmetry in the data.
   - Kurtosis (K) measures how heavy-tailed or light-tailed the distribution is compared to a normal distribution.
2. **Limitations**:
   - The test is sensitive to large sample sizes; even small deviations from normality may result in rejection of H_0.
   - Assumes that the data are independently and identically distributed.

Applications

- Testing normality in regression residuals.
- Validating distributional assumptions in econometrics and time series analysis.

The Jarque-Bera test can be performed in R using the `tseries` package:

```
library(tseries)
#> Registered S3 method overwritten by 'quantmod':
#>   method            from
#>   as.zoo.data.frame zoo

# Generate a sample dataset
data <- rnorm(100)  # Normally distributed data

# Perform the Jarque-Bera test
jarque.bera.test(data)
#>
#>  Jarque Bera Test
#>
#> data:  data
#> X-squared = 1.2822, df = 2, p-value = 0.5267
```

3.4 Bivariate Statistics

Bivariate statistics involve the analysis of relationships between two variables. Understanding these relationships can provide insights into patterns, associations, or (suggestive of) causal connections. Below, we explore the correlation between different types of variables:

- Two Continuous **Variables**
- Two Discrete **Variables**
- Categorical and Continuous **Variables**

Before delving into the analysis, it is critical to consider the following:

1. **Is the relationship linear or non-linear?**
    - Linear relationships can be modeled with simpler statistical methods such as Pearson's correlation, while non-linear relationships may require alternative approaches, such as Spearman's rank correlation or regression with transformations.
2. **If the variable is continuous, is it normal and homoskedastic?**
    - For parametric methods like Pearson's correlation, assumptions such as normality and homoskedasticity (equal variance) must be met. When these assumptions fail, non-parametric methods like Spearman's correlation or robust alternatives are preferred.
3. **How big is your dataset?**
    - Large datasets can reveal subtle patterns but may lead to statistically significant results that are not practically meaningful. For smaller datasets, careful selection of statistical methods is essential to ensure reliability and validity (Table 3.5).

Table 3.5 Common association measures

	Categorical	Continuous
Categorical	Chi-squared test	
	Phi coefficient	
	Cramer's V	
	Tschuprow's T	
	Spearman's rank correlation	
	Kendall's tau	
	Gamma statistic	
	Freeman's theta	
	Epsilon-squared	
	Goodman Kruskal's gamma	
	Somers' D	
	Kendall's Tau-b	
	Yule's Q and Y	
	Tetrachoric correlation	
	Polychoric correlation	
Continuous	Point-Biserial correlation	Pearson correlation
	Logistic regression	Spearman correlation

3.4.1 *Two Continuous*

```
set.seed(1)
n = 100 # (sample size)

data = data.frame(A = sample(1:20, replace = TRUE, size = n),
                  B = sample(1:30, replace = TRUE, size = n))
```

3.4.1.1 Pearson Correlation

Pearson correlation quantifies the strength and direction of a **linear relationship** between two continuous variables.

Formula:

$$r = \frac{\sum(x_i - \bar{x})(y_i - \bar{y})}{\sqrt{\sum(x_i - \bar{x})^2 \cdot \sum(y_i - \bar{y})^2}}$$

where

- x_i, y_i: Individual data points of variables X and Y.
- $\bar{x}$, $\bar{y}$: Means of X and Y.

Assumptions:

1. The relationship between variables is **linear**.
2. Variables are **normally distributed**.
3. Data exhibits **homoscedasticity** (equal variance of Y for all values of X).

Use Case:

- Use when the relationship is expected to be linear, and assumptions of normality and homoscedasticity are met.

Interpretation:

- $r = +1$: Perfect positive linear relationship.
- $r = -1$: Perfect negative linear relationship.
- $r = 0$: No linear relationship.

```
# Pearson correlation
pearson_corr <- stats::cor(data$A, data$B, method = "pearson")
cat("Pearson Correlation (r):", pearson_corr, "\n")
#> Pearson Correlation (r): 0.02394939
```

3.4.1.2 Spearman Correlation

Spearman correlation measures the strength of a **monotonic relationship** between two variables. It ranks the data and calculates correlation based on ranks.

Formula:

$$\rho = 1 - \frac{6 \sum d_i^2}{n(n^2 - 1)}$$

where

- d_i: Difference between the ranks of x_i and y_i.
- n: Number of paired observations.

Assumptions

1. Relationship must be **monotonic**, not necessarily linear.
2. No assumptions about the distribution of variables.

Use Case:

- Use when data is ordinal or when normality and linearity assumptions are violated.

Interpretation:

- $\rho = +1$: Perfect positive monotonic relationship.
- $\rho = -1$: Perfect negative monotonic relationship.

- $\rho = 0$: No monotonic relationship.

```
# Spearman correlation
spearman_corr <- stats::cor(data$A, data$B, method = "spearman")
cat("Spearman Correlation (rho):", spearman_corr, "\n")
#> Spearman Correlation (rho): 0.02304636
```

3.4.1.3 Kendall's Tau Correlation

Kendall's Tau measures the strength of a **monotonic relationship** by comparing concordant and discordant pairs.

Formula:

$$\tau = \frac{(C - D)}{\binom{n}{2}}$$

where

- C: Number of concordant pairs (where ranks of X and Y increase or decrease together).
- D: Number of discordant pairs (where one rank increases while the other decreases).
- $\binom{n}{2}$: Total number of possible pairs.

Assumptions

1. No specific assumptions about the data distribution.
2. Measures monotonic relationships.

Use Case:

- Preferred for small datasets or when data contains outliers.

Interpretation:

- $\tau = +1$: Perfect positive monotonic relationship.
- $\tau = -1$: Perfect negative monotonic relationship.
- $\tau = 0$: No monotonic relationship.

```
# Kendall's Tau correlation
kendall_corr <- stats::cor(data$A, data$B, method = "kendall")
cat("Kendall's Tau Correlation (tau):", kendall_corr, "\n")
#> Kendall's Tau Correlation (tau): 0.02171284
```

3.4.1.4 Distance Correlation

Distance Correlation measures both **linear and non-linear relationships** between variables. It does not require monotonicity or linearity.

Formula:

$$dCor = \frac{dCov(X, Y)}{\sqrt{dVar(X) \cdot dVar(Y)}}$$

where

- $dCov$: Distance covariance between X and Y.
- $dVar$: Distance variances of X and Y.

Assumptions

- No specific assumptions about the relationship (linear, monotonic, or otherwise).

Use Case:

- Use for complex relationships, including non-linear patterns (Table 3.6).

Interpretation:

- $dCor = 0$: No association.
- $dCor = 1$: Perfect association.

```
# Distance correlation
distance_corr <- energy::dcor(data$A, data$B)
cat("Distance Correlation (dCor):", distance_corr, "\n")
#> Distance Correlation (dCor): 0.1008934
```

Table 3.6 Summary of correlation methods

Method	Formula/approach	detects relationship type	Assumptions	Sensitivity to outliers	Use case
Pearson	Linear covariance	Linear	Normality, homoscedasticity	High	Linear relationships
Spearman	Ranks and monotonicity formula	Monotonic	None	Moderate	Monotonic, non-linear data
Kendall's Tau	Concordance/ discordance ratio	Monotonic	None	Low	Small datasets, robust to outliers
Distance correlation	Distance-based variance	Linear and non-linear	None	Low	Complex, non-linear relationships

3.4.2 Categorical and Continuous

Analyzing the relationship between a **categorical variable** (binary or multi-class) and a **continuous variable** requires specialized techniques. These methods assess whether the categorical variable significantly influences the continuous variable or vice versa.

We focus on the following methods:

1. Point-Biserial Correlation
2. Logistic Regression
3. [Analysis of Variance (ANOVA)]
4. T-test

3.4.2.1 Point-Biserial Correlation

The **Point-Biserial Correlation** is a special case of the Pearson correlation used to assess the relationship between a **binary categorical variable** (coded as 0 and 1) and a continuous variable. It measures the strength and direction of the linear relationship.

Formula:

$$r_{pb} = \frac{\bar{Y_1} - \bar{Y_0}}{s_Y} \sqrt{\frac{n_1 n_0}{n^2}}$$

where

- $\bar{Y_1}$, $\bar{Y_0}$: Mean of the continuous variable for the groups coded as 1 and 0, respectively.
- s_Y: Standard deviation of the continuous variable.
- n_1, n_0: Number of observations in each group (1 and 0).
- n: Total number of observations.

Key Properties:

- **Range**: -1 to 1.
  - $r_{pb} = +1$: Perfect positive correlation.
  - $r_{pb} = -1$: Perfect negative correlation.
  - $r_{pb} = 0$: No linear relationship.
- A positive r_{pb} indicates higher values of the continuous variable are associated with the 1 group, while a negative r_{pb} indicates the opposite.

Assumptions

1. The binary variable is **truly dichotomous** (e.g., male/female, success/failure).
2. The continuous variable is approximately **normally distributed**.

3. Homogeneity of variance across the two groups (not strictly required but recommended).

   Use Case:

- To evaluate the linear relationship between a binary categorical variable and a continuous variable.

```
library(ltm)
#> Loading required package: MASS
#>
#> Attaching package: 'MASS'
#> The following object is masked from 'package:EnvStats':
#>
#>     boxcox
#> The following object is masked from 'package:sm':
#>
#>     muscle
#> Loading required package: msm
#> Loading required package: polycor
# Point-Biserial Correlation
biserial_corr <- ltm::biserial.cor(
  c(12.5, 15.3, 10.7, 18.1, 11.2, 16.8, 13.4, 14.9),
  c(0, 1, 0, 1, 0, 1, 0, 1),
  use = "all.obs",
  level = 2
)
cat("Point-Biserial Correlation:", biserial_corr, "\n")
#> Point-Biserial Correlation: 0.8792835
```

3.4.2.2 Logistic Regression

Logistic Regression models the relationship between a **binary categorical variable** (dependent variable) and one or more independent variables (which may include continuous variables). It predicts the probability of the binary outcome (e.g., success/failure, yes/no).

Refer to 3.4.2.2 for more detail.

Formula:

The logistic regression model is represented as:

$$\text{logit}(p) = \ln\left(\frac{p}{1-p}\right) = \beta_0 + \beta_1 X$$

where

- p: Probability of the outcome being 1.
- β_0: Intercept.
- β_1: Coefficient for the continuous variable X.
- $\text{logit}(p)$: Log-odds of the probability.

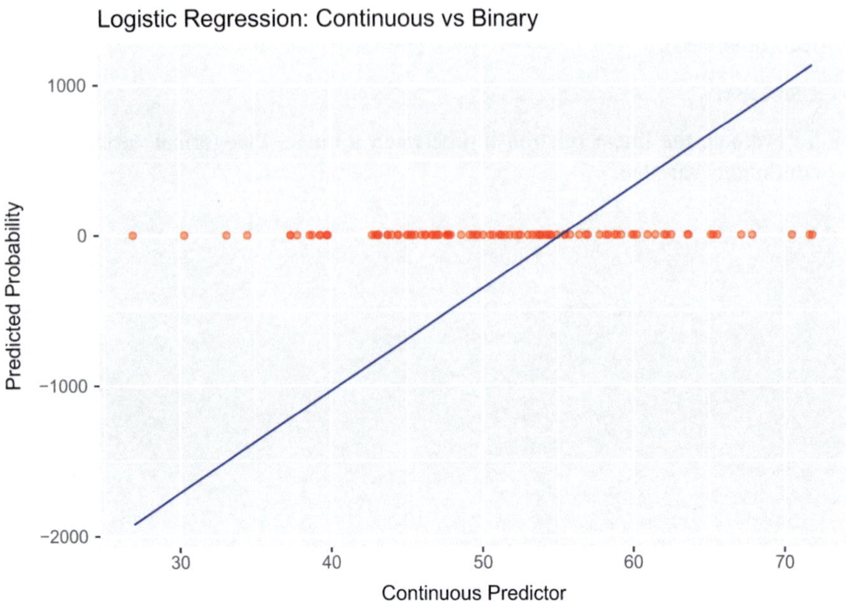

Fig. 3.11 Logistic regression continuous versus binary

Key Features:

- **Output**: Odds ratio or probability of the binary outcome.
- Can include multiple predictors (continuous and categorical).
- Non-linear transformation ensures predictions are probabilities between 0 and 1.

Assumptions

1. The dependent variable is **binary**.
2. Observations are **independent**.
3. There is a **linear relationship** between the logit of the dependent variable and the independent variable.
4. No **multicollinearity** between predictors (Fig. 3.11).

Use Case:

- To predict the likelihood of a binary outcome based on a continuous predictor (e.g., probability of success given test scores) (Table 3.7).

Table 3.7 Summary table of methods (between categorical and continuous)

Method	Type of variable relationship	Key assumptions	Use case
Point-Biserial correlation	Binary categorical vs Continuous	Linear, normality (continuous)	Assess linear association
Logistic regression	Continuous → Binary categorical	Logit-linear relationship	Predict probability of binary outcome
ANOVA	Multi-level categorical vs Continuous	Normality, homogeneity of variance	Compare means across groups
T-test	Binary categorical vs Continuous	Normality, equal variance	Compare means between two groups

```
# Simulated data
set.seed(123)
x <- rnorm(100, mean = 50, sd = 10)  # Continuous predictor
y <- ifelse(x > 55, 1, 0)  # Binary outcome based on threshold

# Logistic Regression
logistic_model <- glm(y ~ x, family = binomial)
#> Warning: glm.fit: algorithm did not converge
#> Warning: glm.fit: fitted probabilities numerically 0 or 1
#> occurred
summary(logistic_model)
#>
#> Call:
#> glm(formula = y ~ x, family = binomial)
#>
#> Coefficients:
#>             Estimate Std. Error z value Pr(>|z|)
#> (Intercept)  -3749.9   495083.0  -0.008    0.994
#> x               67.9     8966.6   0.008    0.994
#>
#> (Dispersion parameter for binomial family taken to be 1)
#>
#>     Null deviance: 1.2217e+02  on 99  degrees of freedom
#> Residual deviance: 1.4317e-07  on 98  degrees of freedom
#> AIC: 4
#>
#> Number of Fisher Scoring iterations: 25

# Predicted probabilities
predicted_probs <- predict(logistic_model, type = "response")
print(head(predicted_probs))
#>         1         2         3         4         5         6
#> -735.6466 -511.3844  703.2134 -307.2281 -267.3187  809.3747
```

```
# Visualize logistic regression curve
library(ggplot2)
data <- data.frame(x = x, y = y, predicted = predicted_probs)

ggplot(data, aes(x = x, y = predicted)) +
    geom_point(aes(y = y), color = "red", alpha = 0.5) +
    geom_line(color = "blue") +
    labs(title = "Logistic Regression: Continuous vs Binary",
         x = "Continuous Predictor", y = "Predicted Probability")
```

3.4.3 Two Discrete

When analyzing the relationship between two **discrete variables** (categorical or ordinal), various methods are available to quantify the degree of association or similarity. These methods can broadly be classified into:

1. Distance Metrics
2. Statistical Metrics

3.4.3.1 Distance Metrics

Distance metrics measure the **dissimilarity** between two discrete variables and are often used as a proxy for correlation in specific applications like clustering or machine learning.

3.4.3.1.1 Euclidean Distance

$$d(x, y) = \sqrt{\sum_{i=1}^{n}(x_i - y_i)^2}$$

- Measures the straight-line distance between two variables in Euclidean space.
- Sensitive to scaling; variables should be normalized for meaningful comparisons.

3.4.3.1.2 Manhattan Distance

$$d(x, y) = \sum_{i=1}^{n} |x_i - y_i|$$

- Measures distance by summing the absolute differences along each dimension.
- Also called **L1 norm**; often used in grid-based problems.

3.4.3.1.3 Chebyshev Distance

$$d(x, y) = \max_{i=1}^{n} |x_i - y_i|$$

- Measures the maximum single-step distance along any dimension.
- Useful in discrete, grid-based problems (e.g., chess moves).

3.4.3.1.4 Minkowski Distance

$$d(x, y) = \left(\sum_{i=1}^{n} |x_i - y_i|^p \right)^{1/p}$$

- Generalized distance metric. Special cases include:
  - $p = 1$: Manhattan Distance.
  - $p = 2$: Euclidean Distance.
  - $p \to \infty$: Chebyshev Distance.

3.4.3.1.5 Canberra Distance

$$d(x, y) = \sum_{i=1}^{n} \frac{|x_i - y_i|}{|x_i| + |y_i|}$$

- Emphasizes proportional differences, making it sensitive to smaller values.

3.4.3.1.6 Hamming Distance

$$d(x, y) = \sum_{i=1}^{n} I(x_i \neq y_i)$$

- Counts the number of differing positions between two sequences.
- Widely used in text similarity and binary data.

3.4.3.1.7 Cosine Similarity and Distance

$$\text{Cosine Similarity} = \frac{\sum_{i=1}^{n} x_i y_i}{\sqrt{\sum_{i=1}^{n} x_i^2} \cdot \sqrt{\sum_{i=1}^{n} y_i^2}}$$

$$\text{Cosine Distance} = 1 - \text{Cosine Similarity}$$

- Measures the angle between two vectors in a high-dimensional space.
- Often used in text and document similarity.

3.4.3.1.8 Sum of Absolute Differences

$$d(x, y) = \sum_{i=1}^{n} |x_i - y_i|$$

- Equivalent to Manhattan Distance but without coordinate context.

3.4.3.1.9 Sum of Squared Differences

$$d(x, y) = \sum_{i=1}^{n} (x_i - y_i)^2$$

- Equivalent to squared Euclidean Distance.

3.4.3.1.10 Mean Absolute Error

$$\text{MAE} = \frac{1}{n} \sum_{i=1}^{n} |x_i - y_i|$$

- Measures average absolute differences.

```r
# Example data
x <- c(1, 2, 3, 4, 5)
y <- c(2, 3, 4, 5, 6)

# Compute distances
euclidean <- sqrt(sum((x - y)^2))
manhattan <- sum(abs(x - y))
chebyshev <- max(abs(x - y))
hamming <- sum(x != y)
cosine_similarity <- sum(x * y) / (sqrt(sum(x^2)) * sqrt(sum(y^2)))
cosine_distance <- 1 - cosine_similarity

# Display results
cat("Euclidean Distance:", euclidean, "\n")
#> Euclidean Distance: 2.236068
cat("Manhattan Distance:", manhattan, "\n")
#> Manhattan Distance: 5
cat("Chebyshev Distance:", chebyshev, "\n")
#> Chebyshev Distance: 1
```

```
cat("Hamming Distance:", hamming, "\n")
#> Hamming Distance: 5
cat("Cosine Distance:", cosine_distance, "\n")
#> Cosine Distance: 0.005063324
```

3.4.3.2 Statistical Metrics

3.4.3.2.1 Chi-squared Test

The **Chi-Squared Test** evaluates whether two categorical variables are **independent** by comparing observed and expected frequencies in a contingency table.

Formula:

$$\chi^2 = \sum \frac{(O_i - E_i)^2}{E_i}$$

where

- O_i: Observed frequency in each cell of the table.
- E_i: Expected frequency under the assumption of independence.

Steps:

1. Construct a contingency table with observed counts.
2. Compute expected frequencies: $E_{ij} = \frac{\text{Row Total}_i \cdot \text{Column Total}_j}{\text{Grand Total}}$
3. Apply the Chi-squared formula.
4. Compare χ^2 with a critical value from the Chi-squared distribution.

Assumptions

1. Observations are **independent**.
2. Expected frequencies should be ≥ 5 in at least 80% of the cells.

Use Case:

- Tests for **independence** between two nominal variables.

```
# Example data
dt <- matrix(c(15, 25, 20, 40), nrow = 2)
rownames(dt) <- c("Group A", "Group B")
colnames(dt) <- c("Category 1", "Category 2")

# Perform Chi-Squared Test
chi_sq_test <- chisq.test(dt)
print(chi_sq_test)
#>
#> Results of Hypothesis Test
#> --------------------------
#>
#> Alternative Hypothesis:
#>
#> Test Name:                       Pearson's Chi-squared test with Yates' continuity correction
```

```
#>
#> Data:                          dt
#>
#> Test Statistic:                X-squared = 0.04578755
#>
#> Test Statistic Parameter:      df = 1
#>
#> P-value:                       0.8305625
```

3.4.3.2.2 Phi Coefficient

The **Phi Coefficient** is a measure of association between two **binary variables**, derived from the Chi-squared statistic.

Formula:

$$\phi = \frac{\chi^2}{n}$$

where

- n: Total sample size.

Interpretation:

- $\phi = 0$: No association.
- $\phi = +1$: Perfect positive association.
- $\phi = -1$: Perfect negative association.

Use Case:

- Suitable for **2×2 contingency tables**.
- 2 binary

```
library(psych)
#>
#> Attaching package: 'psych'
#> The following object is masked from 'package:ltm':
#>
#>     factor.scores
#> The following object is masked from 'package:polycor':
#>
#>     polyserial
#> The following objects are masked from 'package:DescTools':
#>
#>     AUC, ICC, SD
#> The following objects are masked from 'package:ggplot2':
#>
#>     %+%, alpha
#> The following object is masked from 'package:car':
#>
#>     logit
```

```
# Compute Phi Coefficient
phi_coeff <- phi(dt)
cat("Phi Coefficient:", phi_coeff, "\n")
#> Phi Coefficient: 0.04
```

3.4.3.2.3 Cramer's V

Cramer's V generalizes the Phi coefficient to handle contingency tables with more than two rows or columns.

Formula:

$$V = \sqrt{\frac{\chi^2/n}{\min(r-1, c-1)}}$$

where

- r: Number of rows.
- c: Number of columns.

Assumptions

1. Variables are nominal.
2. Suitable for **larger contingency tables**.

Use Case:

- Measures the **strength of association** between nominal variables with no natural order.

```
library(lsr)
#>
#> Attaching package: 'lsr'
#> The following object is masked from 'package:EnvStats':
#>
#>     longToWide

# Simulate data
set.seed(1)
data <- data.frame(
  A = sample(1:5, replace = TRUE, size = 100),  # Nominal variable
  B = sample(1:6, replace = TRUE, size = 100)  # Nominal variable
)

# Compute Cramer's V
cramers_v <- cramersV(data$A, data$B)
#> Warning in stats::chisq.test(...): Chi-squared
#> approximation may be incorrect
cat("Cramer's V:", cramers_v, "\n")
#> Cramer's V: 0.1944616
```

Alternatively,

- `ncchisq` noncentral Chi-square
- `nchisqadj` Adjusted noncentral Chi-square
- `fisher` Fisher Z transformation
- `fisheradj` bias correction Fisher z transformation

```
DescTools::CramerV(data, conf.level = 0.95,method = "ncchisqadj")
#>  Cramer V    lwr.ci    upr.ci
#> 0.3472325 0.3929964 0.4033053
```

3.4.3.2.4 Adjusted Cramer's V

Adjusted versions of Cramer's V correct for bias, especially in small samples.

Adjusted formulas account for non-central Chi-squared or bias correction. Examples include:

- **Non-central Chi-squared**: $V_{adj} = \sqrt{\frac{\chi^2_{nc}/n}{\min(r-1,c-1)}}$
- **Bias Correction**: $V_{adj} = V - \text{Bias Term}$

```
library(DescTools)

# Compute Adjusted Cramer's V
cramers_v_adj <- CramerV(data, conf.level = 0.95, method = "ncchisqadj")
cramers_v_adj
#>  Cramer V    lwr.ci    upr.ci
#> 0.3472325 0.3929964 0.4033053
```

3.4.3.2.5 Tschuprow's T

Tschuprow's T is a symmetric measure of association for nominal variables. It differs from Cramer's V by considering the product of rows and columns, making it less sensitive to asymmetrical tables.

Formula:

$$T = \sqrt{\frac{\chi^2/n}{\sqrt{(r-1)(c-1)}}}$$

Assumptions

- Applicable to nominal variables.
- Suitable for contingency tables with unequal dimensions.

Use Case:

- Preferred when table dimensions are **highly unequal**.

```
# Compute Tschuprow's T
tschuprow_t <- DescTools::TschuprowT(data$A, data$B)
cat("Tschuprow's T:", tschuprow_t, "\n")
#> Tschuprow's T: 0.1839104
```

3.4.3.2.6 Ordinal Association (Rank correlation)

When at least one variable is **ordinal**, rank-based methods are the most appropriate as they respect the **order of the categories**. These methods are often used when relationships are monotonic (increasing or decreasing consistently) but not necessarily linear.

Spearman's Rank Correlation

Spearman's Rank Correlation (ρ) measures the strength and direction of a **monotonic relationship** between two variables. It transforms the data into ranks and calculates Pearson correlation on the ranks.

Formula:

$$\rho = 1 - \frac{6 \sum d_i^2}{n(n^2 - 1)}$$

where

- d_i: Difference between the ranks of the paired observations.
- n: Number of paired observations.

Assumptions

1. Data must be **ordinal** or **continuous** but convertible to ranks.
2. Relationship is **monotonic**.

Use Case:

- Suitable for ordinal-ordinal or ordinal-continuous associations.

```
# Simulating ordinal data
set.seed(123)
ordinal_x <- sample(1:5, 100, replace = TRUE)
ordinal_y <- sample(1:5, 100, replace = TRUE)

# Spearman's Correlation
spearman_corr <- cor(ordinal_x, ordinal_y, method = "spearman")
cat("Spearman's Correlation (rho):", spearman_corr, "\n")
#> Spearman's Correlation (rho): 0.08731195
```

Kendall's Tau

Kendall's Tau (τ) measures the strength of a **monotonic relationship** by comparing concordant and discordant pairs.

Formula:

$$\tau = \frac{C - D}{C + D}$$

where

- C: Number of concordant pairs (ranks increase together).
- D: Number of discordant pairs (one rank increases while the other decreases).

Variants:

1. **Kendall's Tau-a**: For data with no ties.
2. **Kendall's Tau-b**: Adjusted for ties in ranks.
3. **Kendall's Tau-c**: Adjusted for ties in large tables.

Use Case:

- Ideal for small datasets or when ties are present.

```
# Kendall's Tau
kendall_corr <- cor(ordinal_x, ordinal_y, method = "kendall")
cat("Kendall's Tau (tau):", kendall_corr, "\n")
#> Kendall's Tau (tau): 0.06795076
```

Gamma Statistic

The **Gamma Statistic** measures the strength of association between two ordinal variables by focusing on **concordant** and **discordant** pairs, ignoring ties.

Formula:

$$\gamma = \frac{C - D}{C + D}$$

Use Case:

- Works well when there are **many ties** in the data.

```
library(vcd)
#> Loading required package: grid

# Simulating ordinal data
cont_table <- table(ordinal_x, ordinal_y)

# Gamma Statistic
gamma_stat <- assocstats(cont_table)$gamma
cat("Gamma Statistic:", gamma_stat, "\n")
#> Gamma Statistic:
```

Freeman's Theta

Freeman's Theta measures the association between an **ordinal variable** and a **nominal variable**. It quantifies how well the grouping in the nominal variable explains the ordering in the ordinal variable.

Use Case:

- Useful when analyzing relationships between ordinal predictors and nominal responses (or vice versa).

```
rcompanion::freemanTheta(ordinal_x, ordinal_y)
#> Freeman.theta
#>          0.094
```

Epsilon-squared

Epsilon-Squared (ϵ^2) measures the proportion of variance in the **ordinal variable** explained by a **nominal variable**. It is conceptually similar to the coefficient of determination (R^2) in linear regression but adapted for ordinal-nominal relationships.

Formula:

$$\epsilon^2 = \frac{\text{variance explained by group differences}}{\text{total variance}}$$

where

- The numerator represents the variance between ordinal categories due to differences in nominal groups.
- The denominator is the total variance in the ordinal variable.

Use Case:

- Quantifies the effect size when analyzing how well a nominal variable explains an ordinal variable.

```
set.seed(123)
ordinal_x <- sample(1:5, 100, replace = TRUE)  # Ordinal variable
nominal_y <- sample(1:3, 100, replace = TRUE) # Nominal variable

# Compute Epsilon-Squared
library(rcompanion)
#>
#> Attaching package: 'rcompanion'
#> The following object is masked from 'package:psych':
#>
#>     phi
epsilon_squared <- rcompanion::epsilonSquared(ordinal_x, nominal_y)
print(epsilon_squared)
#> epsilon.squared
#>          0.00446
```

Goodman-Kruskal's Gamma

Goodman-Kruskal's Gamma measures the strength of association between two **ordinal variables**. It is a rank-based measure, focusing only on **concordant** and **discordant** pairs while ignoring ties.

Formula:

$$\gamma = \frac{C - D}{C + D}$$

where

- C: Number of concordant pairs (where ranks move in the same direction).
- D: Number of discordant pairs (where ranks move in opposite directions).

Use Case:

- Suitable for ordinal variables with many ties.

```
n = 100 # (sample size)
set.seed(1)
dt = table(data.frame(
    A = sample(1:4, replace = TRUE, size = n), # ordinal
    B = sample(1:3, replace = TRUE, size = n)  # ordinal
))
dt
#>    B
#> A   1  2  3
#>   1  7 11  9
#>   2 11  6 14
#>   3  7 11  4
#>   4  6  4 10
```

```
# Compute Goodman-Kruskal's Gamma
library(DescTools)
goodman_kruskal_gamma <- GoodmanKruskalGamma(dt, conf.level = 0.95)
cat("Goodman-Kruskal's Gamma:", goodman_kruskal_gamma, "\n")
#> Goodman-Kruskal's Gamma: 0.006781013 -0.2290321 0.2425941
```

Somers' D

Somers' D (also called Somers' Delta) extends Kendall's Tau by focusing on **asymmetric relationships**, where one variable is a predictor and the other is a response.

Formula:

$$D_{XY} = \frac{C - D}{C + D + T_Y}$$

where

- T_Y: Tied pairs in the dependent variable.

Use Case:

- Appropriate when there is a clear predictor-response relationship between two ordinal variables.

```
# Compute Somers' D
somers_d <- SomersDelta(dt, conf.level = 0.95)
somers_d
#>        somers       lwr.ci       upr.ci
#>   0.005115859 -0.172800185  0.183031903
```

Kendall's Tau-b

Kendall's Tau-b is an extension of Kendall's Tau that accounts for **ties** in the data.

Formula:

$$\tau_b = \frac{C - D}{\sqrt{(C + D + T_X)(C + D + T_Y)}}$$

where

- T_X, T_Y: Tied pairs in each variable.

Use Case:

- Use when ordinal data contains ties.

```
# Compute Kendall's Tau-b
kendalls_tau_b <- KendallTauB(dt, conf.level = 0.95)
kendalls_tau_b
#>         tau_b       lwr.ci       upr.ci
#>   0.004839732 -0.163472443  0.173151906
```

Yule's Q and Y

Yule's Q and **Yule's Y** are specialized measures for **2×2 contingency tables**. They are simplified versions of Goodman-Kruskal's Gamma, designed for binary ordinal variables.

Use Case:

- Ideal for binary ordinal variables in a 2×2 table.

Special version (2×2) of the Goodman Kruskal's Gamma coefficient (Table 3.8).

$$\text{Yule's Q} = \frac{ad - bc}{ad + bc}$$

Table 3.8 Special version of the Goodman Kruskal's gamma coefficient

	Variable 1	
Variable 2	a	b
	c	d

We typically use Yule's Q in practice while Yule's Y has the following relationship with Q.

$$\text{Yule's Y} = \frac{\sqrt{ad} - \sqrt{bc}}{\sqrt{ad} + \sqrt{bc}}$$

$$Q = \frac{2Y}{1 + Y^2}$$

$$Y = \frac{1 = \sqrt{1 - Q^2}}{Q}$$

```
# Create 2x2 table
dt_binary <- table(data.frame(
  A = sample(c(0, 1), replace = TRUE, size = n),
  B = sample(c(0, 1), replace = TRUE, size = n)
))

# Compute Yule's Q
yules_q <- YuleQ(dt_binary)
yules_q
#> [1] -0.07667474
```

Tetrachoric Correlation

Tetrachoric Correlation measures the association between two **binary variables** by assuming they represent thresholds of underlying continuous normal distributions. It is a special case of Polychoric Correlation when both variables are binary

```
# Simulate binary data
library(psych)
data_binary <- data.frame(
  A = sample(c(0, 1), replace = TRUE, size = n),
  B = sample(c(0, 1), replace = TRUE, size = n)
)

# Compute Tetrachoric Correlation
tetrachoric_corr <- tetrachoric(data_binary)
print(tetrachoric_corr)
#> $rho
#>            A          B
```

```
#> A 1.0000000 0.3147214
#> B 0.3147214 1.0000000
#>
#> $tau
#>            A           B
#>  0.12566135 -0.02506891
#>
#> $n.obs
#> [1] 100
#>
#> $Call
#> tetrachoric(x = data_binary)
#>
#> attr(,"class")
#> [1] "psych" "tetra"
```

Polychoric Correlation

Polychoric Correlation measures the association between **ordinal variables** by assuming they are discretized versions of latent, normally distributed continuous variables (Fig. 3.12).

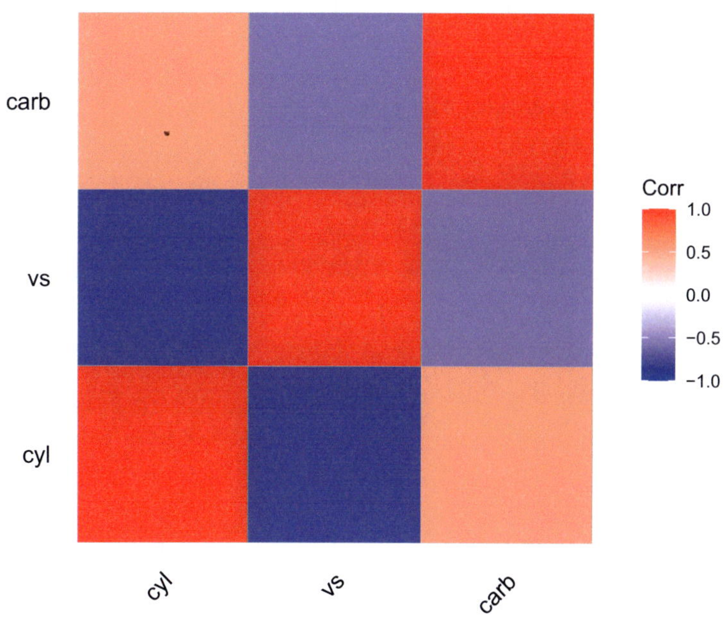

Fig. 3.12 Heatmap example

Table 3.9 Association metrics, variable types, and use cases

Metric	Variable types	Use case
Spearman's correlation	Ordinal vs. Ordinal	Non-linear, monotonic relationships
Kendall's tau	Ordinal vs. Ordinal	Non-linear, monotonic relationships with ties
Gamma statistic	Ordinal vs. Ordinal	Handles data with many ties effectively
Freeman's theta	Ordinal vs. Nominal	Mixed data types (ordinal and nominal)
Epsilon-squared	Ordinal vs. Nominal	Variance explained by nominal groups
Goodman-Kruskal's gamma	Ordinal vs. Ordinal	Strong association; ignores ties
Somers' D	Ordinal predictor and response	Asymmetric association
Kendall's tau-b	Ordinal vs. Ordinal	Adjusts for ties in data
Yule's Q	Binary ordinal vs. Binary ordinal	Special case for 2×2 tables
Tetrachoric correlation	Binary vs. Binary	Binary ordinal variables
Polychoric correlation	Ordinal vs. Ordinal	Continuous latent structure

Assumptions

- The ordinal variables represent categories of an underlying normal distribution.

  Use Case:

- Suitable for ordinal variables with a natural order (Table 3.9).

```
# Simulate ordinal data
library(polycor)
data_ordinal <- data.frame(
  A = sample(1:4, replace = TRUE, size = n),
  B = sample(1:6, replace = TRUE, size = n)
)

# Compute Polychoric Correlation
polychoric_corr <- polychor(data_ordinal$A, data_ordinal$B)
cat("Polychoric Correlation:", polychoric_corr, "\n")
#> Polychoric Correlation: 0.1908334
```

3.4.4 General Approach to Bivariate Statistics

```
library(tidyverse)
#> -- Attaching core tidyverse packages ---- tidyverse 2.0.0 --
#> v dplyr     1.1.4     v readr     2.1.5
#> v forcats   1.0.0     v stringr   1.5.1
```

```
#> v lubridate 1.9.4     v tibble    3.2.1
#> v purrr     1.0.4     v tidyr     1.3.1
#> -- Conflicts ---------------------- tidyverse_conflicts() --
#> x psych::%+%()    masks ggplot2::%+%()
#> x psych::alpha()  masks ggplot2::alpha()
#> x dplyr::filter() masks stats::filter()
#> x dplyr::lag()    masks stats::lag()
#> x dplyr::recode() masks car::recode()
#> x dplyr::select() masks MASS::select()
#> x purrr::some()   masks car::some()
#> i Use the conflicted package (<http://conflicted.r-lib.org/>) to force all conflicts to become errors

data("mtcars")
df = mtcars %>%
    dplyr::select(cyl, vs, carb)


df_factor = df %>%
    dplyr::mutate(
        cyl = factor(cyl),
        vs = factor(vs),
        carb = factor(carb)
    )
# summary(df)
str(df)
#> 'data.frame':    32 obs. of  3 variables:
#>  $ cyl : num  6 6 4 6 8 6 8 4 4 6 ...
#>  $ vs  : num  0 0 1 1 0 1 0 1 1 1 ...
#>  $ carb: num  4 4 1 1 2 1 4 2 2 4 ...
str(df_factor)
#> 'data.frame':    32 obs. of  3 variables:
#>  $ cyl : Factor w/ 3 levels "4","6","8": 2 2 1 2 3 2 3 1 1 2 ...
#>  $ vs  : Factor w/ 2 levels "0","1": 1 1 2 2 1 2 1 2 2 2 ...
#>  $ carb: Factor w/ 6 levels "1","2","3","4",..: 4 4 1 1 2 1 4 2 2 4 ...
```

Get the correlation table for continuous variables only

```
cor(df)
#>             cyl         vs       carb
#> cyl   1.0000000 -0.8108118  0.5269883
#> vs   -0.8108118  1.0000000 -0.5696071
#> carb  0.5269883 -0.5696071  1.0000000

# only complete obs
# cor(df, use = "complete.obs")
```

Alternatively, you can also have the

```
Hmisc::rcorr(as.matrix(df), type = "pearson")
#>        cyl    vs  carb
#> cyl   1.00 -0.81  0.53
#> vs   -0.81  1.00 -0.57
#> carb  0.53 -0.57  1.00
#>
#> n= 32
#>
#>
#> P
#>      cyl    vs     carb
#> cyl         0.0000 0.0019
```

```
#> vs   0.0000          0.0007
#> carb 0.0019 0.0007
```

```
modelsummary::datasummary_correlation(df)
```

```
ggcorrplot::ggcorrplot(cor(df))
```

	cyl	vs	carb
cyl	1	.	.
vs	−0.81	1	.
carb	0.53	−0.57	1

Comparing correlations between different types of variables (e.g., continuous vs. categorical) poses unique challenges. One key issue is ensuring that methods are appropriate for the nature of the variables being analyzed. Another challenge lies in detecting non-linear relationships, as traditional correlation measures, such as Pearson's correlation coefficient, are designed to assess linear associations.

To address these challenges, a potential solution is to utilize **mutual information** from information theory. Mutual information quantifies how much knowing one variable reduces the uncertainty of another, providing a more general measure of association that accommodates both linear and non-linear relationships.

3.4.4.1 Approximating Mutual Information

We can approximate mutual information using the following relationship:

$$\downarrow \text{Prediction Error} \approx \downarrow \text{Uncertainty} \approx \uparrow \text{Association Strength}$$

This principle underpins the X2Y metric, which is implemented through the following steps:

1. **Predict** y without x (baseline model):
   - If y is continuous, predict the mean of y.
   - If y is categorical, predict the mode of y.
2. **Predict** y with x using a model (e.g., linear regression, random forest, etc.).
3. **Calculate the difference in prediction error** between steps 1 and 2. This difference reflects the reduction in uncertainty about y when x is included, serving as a measure of association strength.

3.4.4.2 Generalizing Across Variable Types

To construct a comprehensive framework that handles different variable combinations, such as:

- Continuous vs. continuous
- Categorical vs. continuous
- Continuous vs. categorical
- Categorical vs. categorical

a flexible modeling approach is required. **Classification and Regression Trees (CART)** are particularly well-suited for this purpose, as they can accommodate both continuous and categorical predictors and outcomes. However, other models, such as random forests or generalized additive models (GAMs), may also be employed.

3.4.4.3 Limitations of the Approach

Despite its strengths, this approach has some limitations:

1. **Asymmetry:**
   The measure is not symmetric, meaning $(x, y) \neq (y, x)$.
2. **Comparability:**
   Different variable pairs may yield metrics that are not directly comparable. For instance, continuous outcomes often use metrics like Mean Absolute Error (MAE), while categorical outcomes use measures like misclassification error.

These limitations should be considered when interpreting results, especially in multi-variable or mixed-data contexts (Figs. 3.13, 3.14, 3.15, 3.16, 3.17, 3.18, and 3.19).

```
library(ppsr)
library(tidyverse)

iris <- iris %>%
  dplyr::select(1:3)

# ppsr::score_df(iris) # if you want a dataframe
ppsr::score_matrix(iris,
                   do_parallel = TRUE,
                   n_cores = parallel::detectCores() / 2)
#>              Sepal.Length Sepal.Width Petal.Length
#> Sepal.Length   1.00000000  0.04632352    0.5491398
#> Sepal.Width    0.06790301  1.00000000    0.2376991
#> Petal.Length   0.61608360  0.24263851    1.0000000

# if you want a similar correlation matrix
ppsr::score_matrix(df,
                   do_parallel = TRUE,
                   n_cores = parallel::detectCores() / 2)
```

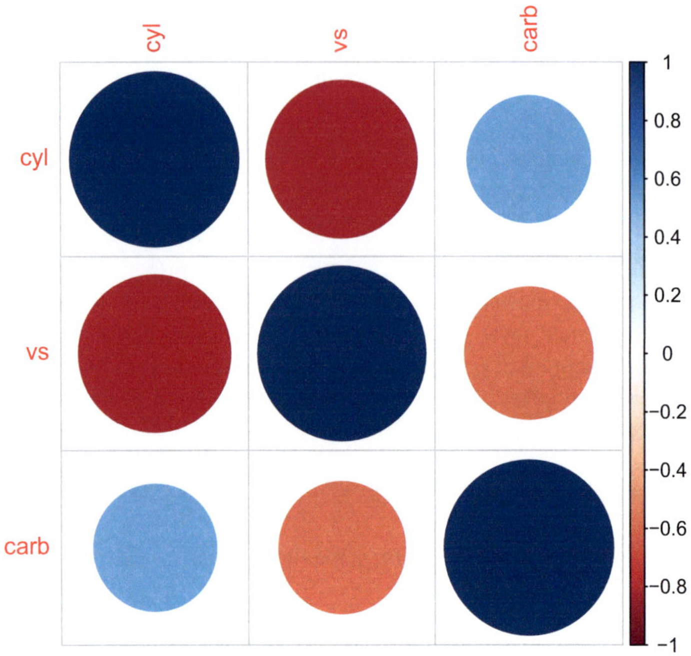

Fig. 3.13 Bubble chart example

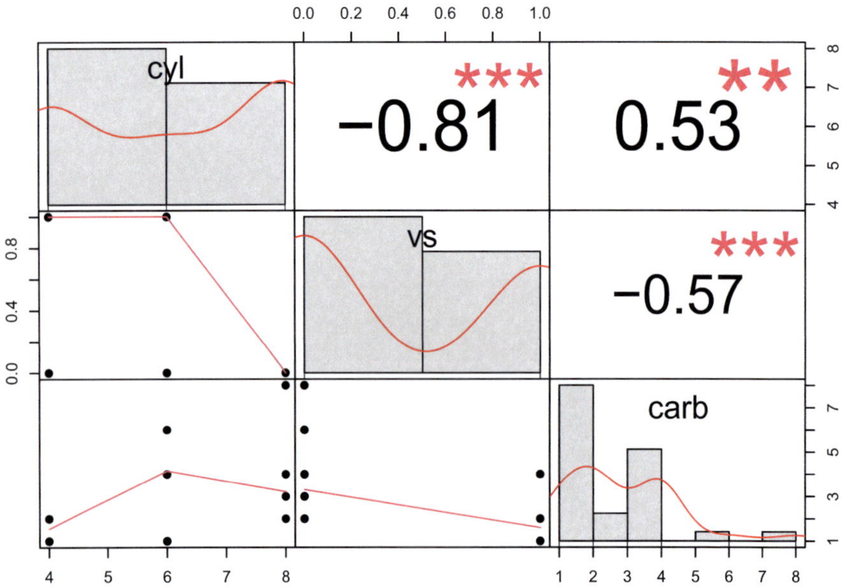

Fig. 3.14 Scatterplot matrix example

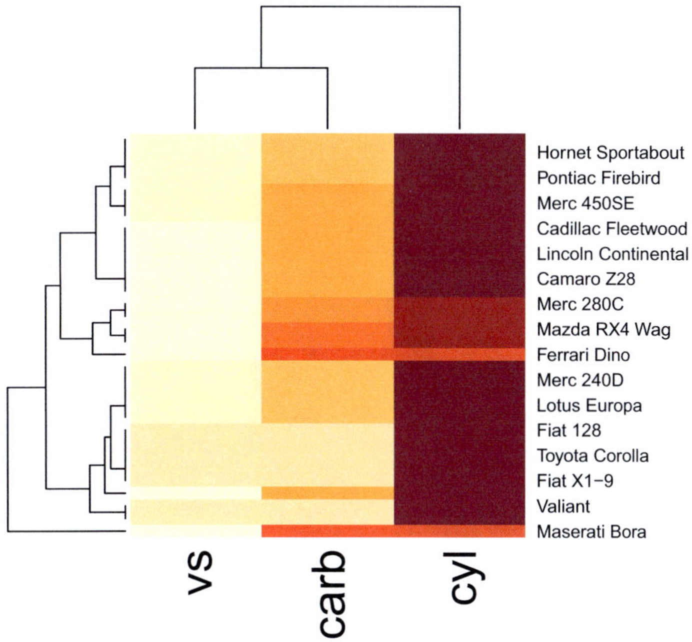

Fig. 3.15 Hierarchical heatmap example

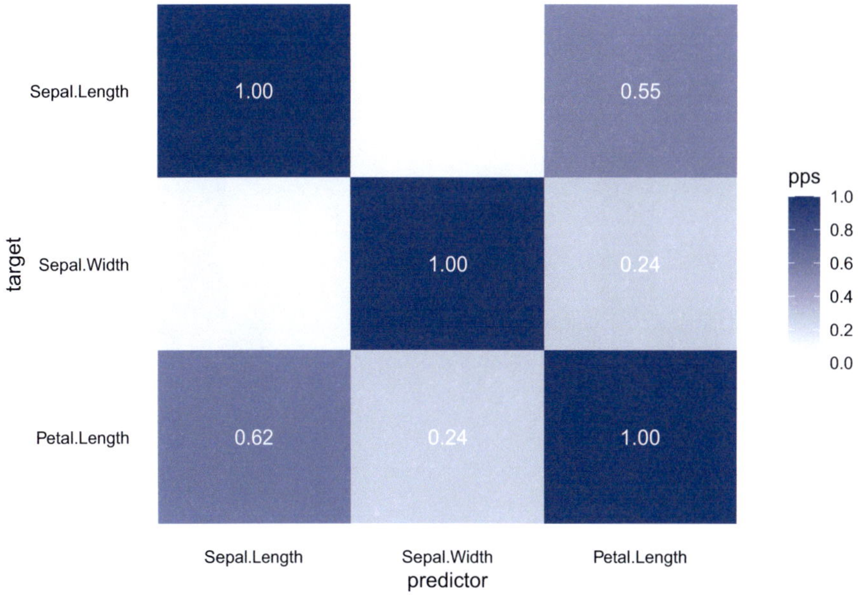

Fig. 3.16 Heatmap predictive example

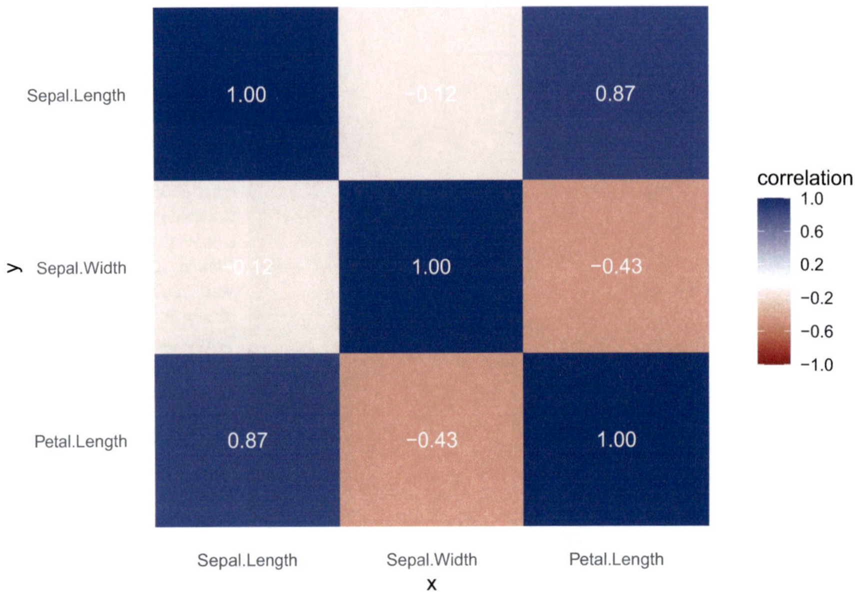

Fig. 3.17 Heatmap correlation example

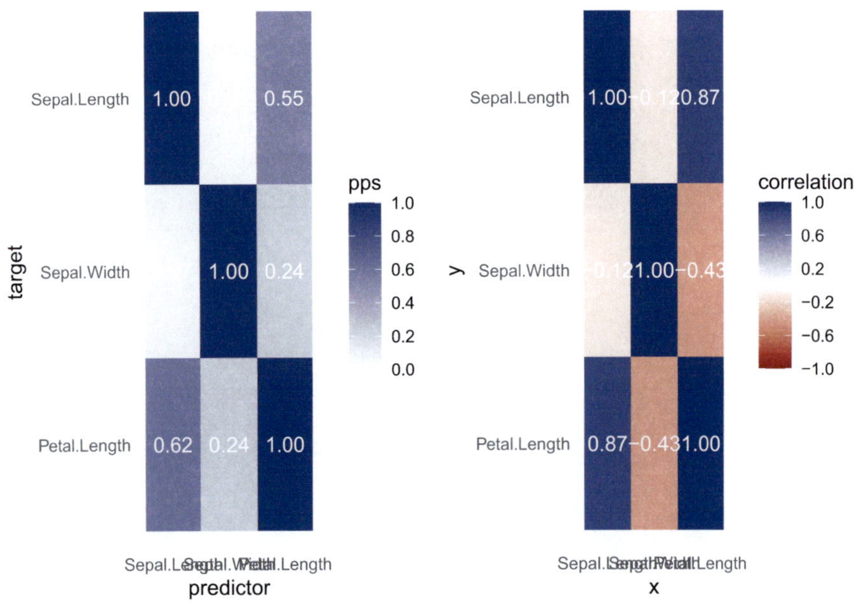

Fig. 3.18 Heatmap and correlation example

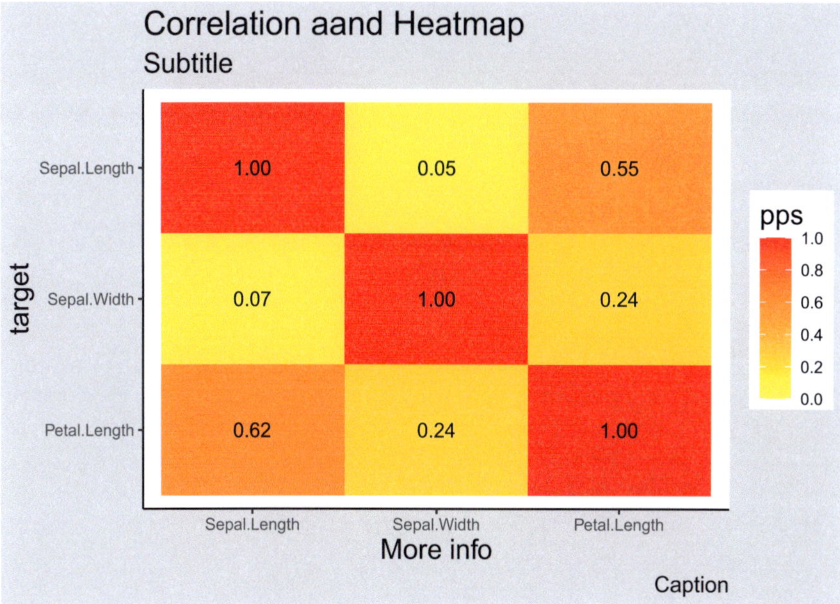

Fig. 3.19 Correlation and heatmap example

```
#>             cyl        vs      carb
#> cyl  1.00000000 0.3982789 0.2092533
#> vs   0.02514286 1.0000000 0.2000000
#> carb 0.30798148 0.2537309 1.0000000
```

```
corrplot::corrplot(cor(df))
```

Alternatively,

```
PerformanceAnalytics::chart.Correlation(df, histogram = T, pch = 19)
```

```
heatmap(as.matrix(df))
```

More general form,

```
ppsr::visualize_pps(
    df = iris,
    do_parallel = TRUE,
    n_cores = parallel::detectCores() / 2
)
```

```
ppsr::visualize_correlations(
    df = iris
)
```

Both heat map and correlation at the same time

```
ppsr::visualize_both(
    df = iris,
    do_parallel = TRUE,
    n_cores = parallel::detectCores() / 2
)
```

More elaboration with `ggplot2`

```
ppsr::visualize_pps(
    df = iris,
    color_value_high = 'red',
    color_value_low = 'yellow',
    color_text = 'black'
) +
    ggplot2::theme_classic() +
    ggplot2::theme(plot.background =
                       ggplot2::element_rect(fill = "lightgrey")) +
    ggplot2::theme(title = ggplot2::element_text(size = 15)) +
    ggplot2::labs(
        title = 'Correlation aand Heatmap',
        subtitle = 'Subtitle',
        caption = 'Caption',
        x = 'More info'
    )
```

Chapter 4
Basic Statistical Inference

This chapter introduces the core logic of statistical inference, focusing on how we use sample data to make probabilistic statements about populations. We begin with the hypothesis testing framework, including null and alternative hypotheses, test statistics, and p-values. Key concepts such as confidence intervals and type I/II errors are explained in depth. Specific methods for one-sample and two-sample inferences are discussed, along with inference for categorical data. We also introduce divergence metrics, which provide alternative ways to compare distributions—a concept increasingly important in modern data science applications.

Statistical inference involves drawing conclusions about population parameters based on sample data. The two primary goals of inference are:

1. *Making inferences* about the true parameter value (β) based on our estimator or estimate:
    - This involves interpreting the sample-derived estimate to understand the population parameter.
    - Examples include estimating population means, variances, or proportions.
2. *Testing whether underlying assumptions hold true*, including:
    - Assumptions about the true population parameters (e.g., μ, σ^2)
    - Assumptions about random variables (e.g., independence, normality)
    - Assumptions about the model specification (e.g., linearity in regression)

Note Statistical testing does not:

- Confirm with absolute certainty that a hypothesis is true or false.
- Interpret the magnitude of the estimated value in economic, practical, or business contexts without additional analysis.

M. Nguyen, *Foundations of Data Analysis*,
https://doi.org/10.1007/978-3-032-01858-8_4

- **Statistical significance**: Refers to whether an observed effect is unlikely due to chance.
- **Practical significance**: Focuses on the real-world importance of the effect.

Example

- A marketing campaign increases sales by 0.5%, which is statistically significant ($p < 0.05$). However, in a small market, this may lack practical significance.

Instead, inference provides a framework for making probabilistic statements about population parameters, given sample data.

4.1 Hypothesis Testing Framework

Hypothesis testing is one of the fundamental tools in statistics. It provides a formal procedure to test claims or assumptions (hypotheses) about population parameters using sample data. This process is essential in various fields, including business, medicine, and social sciences, as it helps answer questions like "Does a new marketing strategy improve sales?" or "Is there a significant difference in test scores between two teaching methods?"

The goal of hypothesis testing is to make decisions or draw conclusions about a population based on sample data. This is necessary because we rarely have access to the entire population. For example, if a company wants to determine whether a new advertising campaign increases sales, it might analyze data from a sample of stores rather than every store globally.

Key Steps in Hypothesis Testing

1. **Formulate Hypotheses**: Define the null and alternative hypotheses.
2. **Choose a Significance Level** (α): Determine the acceptable probability of making a type I error.
3. **Select a Test Statistic**: Identify the appropriate statistical test based on the data and hypotheses.
4. **Define the Rejection Region**: Specify the range of values for which the null hypothesis will be rejected.
5. **Compute the Test Statistic**: Use sample data to calculate the test statistic.
6. **Make a Decision**: Compare the test statistic to the critical value or use the p-value to decide whether to reject or fail to reject the null hypothesis.

4.1.1 Null and Alternative Hypotheses

At the heart of hypothesis testing lies the formulation of two competing hypotheses:

1. **Null Hypothesis** (H_0):
    - Represents the current state of knowledge, status quo, or no effect.
    - It is assumed true unless there is strong evidence against it.
    - Examples:
        - $H_0 : \mu_1 = \mu_2$ (no difference in means between two groups).
        - $H_0 : \beta = 0$ (a predictor variable has no effect in a regression model).
    - Think of H_0 as the "default assumption."
2. **Alternative Hypothesis** (H_a or H_1):
    - Represents a claim that contradicts the null hypothesis.
    - It is what you are trying to prove or find evidence for.
    - Examples:
        - $H_a : \mu_1 \neq \mu_2$ (means of two groups are different).
        - $H_a : \beta \neq 0$ (a predictor variable has an effect).

4.1.2 Errors in Hypothesis Testing

Hypothesis testing involves decision-making under uncertainty, meaning there is always a risk of making errors. These errors are classified into two types:

1. **Type I Error** (α):
    - Occurs when the null hypothesis is rejected, even though it is true.
    - Example: Concluding that a medication is effective when it actually has no effect.
    - The probability of making a type I error is denoted by α, called the *significance level* (commonly set at 0.05 or 5%).
2. **Type II Error** (β):
    - Occurs when the null hypothesis is not rejected, but the alternative hypothesis is true.
    - Example: Failing to detect that a medication is effective when it actually works.
    - The complement of β is called the *power* of the test $(1 - \beta)$, representing the probability of correctly rejecting the null hypothesis.

Analogy: The Legal System
To make this concept more intuitive, consider the analogy of a courtroom:

- **Null Hypothesis** (H_0): The defendant is innocent.
- **Alternative Hypothesis** (H_a): The defendant is guilty.

- **Type I Error**: Convicting an innocent person (false positive).
- **Type II Error**: Letting a guilty person go free (false negative).

Balancing α and β is critical in hypothesis testing, as reducing one often increases the other. For example, if you make it harder to reject H_0 (reducing α), you increase the chance of failing to detect a true effect (increasing β).

4.1.3 *The Role of Distributions in Hypothesis Testing*

Distributions play a fundamental role in hypothesis testing because they provide a mathematical model for understanding how a test statistic behaves under the null hypothesis (H_0). Without distributions, it would be impossible to determine whether the observed results are due to random chance or provide evidence to reject the null hypothesis.

4.1.3.1 Expected Outcomes

One of the key reasons distributions are so crucial is that they describe the range of values a test statistic is likely to take when H_0 is true. This helps us understand what is considered "normal" variation in the data due to random chance. For example:

- Imagine you are conducting a study to test whether a new marketing strategy increases the average monthly sales. Under the null hypothesis, you assume the new strategy has no effect, and the average sales remain unchanged.
- When you collect a sample and calculate the test statistic, you compare it to the expected distribution (e.g., the normal distribution for a z-test). This distribution shows the range of test statistic values that are likely to occur purely due to random fluctuations in the data, assuming H_0 is true.

By providing this baseline of what is "normal," distributions allow us to identify unusual results that may indicate the null hypothesis is false.

4.1.3.2 Critical Values and Rejection Regions

Distributions also help define critical values and rejection regions in hypothesis testing. Critical values are specific points on the distribution that mark the boundaries of the rejection region. The rejection region is the range of values for the test statistic that lead us to reject H_0.

The location of these critical values depends on:

- The *level of significance* (α), which is the probability of rejecting H_0 when it is true (a type I error)
- The shape of the test statistic's distribution under H_0

For example:

- In a one-tailed z-test with $\alpha = 0.05$, the critical value is approximately 1.645 for a standard normal distribution. If the calculated test statistic exceeds this value, we reject H_0 because such a result would be very unlikely under H_0.

Distributions help us visually and mathematically determine these critical points. By examining the distribution, we can see where the rejection region lies and what the probability is of observing a value in that region by random chance alone.

4.1.3.3 P-values

The p-value, a central concept in hypothesis testing, is directly derived from the distribution of the test statistic under H_0. The p-value represents the probability of observing a test statistic as extreme as (or more extreme than) the one calculated, assuming H_0 is true.

The *p-value* quantifies the strength of evidence against H_0. It represents the probability of observing a test statistic as extreme as (or more extreme than) the one calculated, assuming H_0 is true:

- **Small p-value** ($< \alpha$): Strong evidence against H_0; reject H_0.
- **Large p-value** ($> \alpha$): Weak evidence against H_0; fail to reject H_0.

For example:

- Suppose you calculate a z-test statistic of 2.1 in a one-tailed test. Using the standard normal distribution, the p-value is the area under the curve to the right of $z = 2.1$. This area represents the likelihood of observing a result as extreme as $z = 2.1$ if H_0 is true.
- In this case, the p-value is approximately 0.0179. A small p-value (typically less than $\alpha = 0.05$) suggests that the observed result is unlikely under H_0 and provides evidence to reject the null hypothesis.

4.1.3.4 Why Does All This Matter?

To summarize, distributions are the backbone of hypothesis testing because they allow us to:

- Define what is expected under H_0 by modeling the behavior of the test statistic.
- Identify results that are unlikely to occur by random chance, which leads to the rejection of H_0.
- Calculate p-values to quantify the strength of evidence against H_0.

Distributions provide the framework for understanding the role of chance in statistical analysis. They are essential for determining expected outcomes, setting thresholds for decision-making (critical values and rejection regions), and calculating p-values. A solid grasp of distributions will greatly enhance your ability to interpret and conduct hypothesis tests, making it easier to draw meaningful conclusions from data.

4.1.4 The Test Statistic

The test statistic is a crucial component in hypothesis testing, as it quantifies how far the observed data deviates from what we would expect if the null hypothesis (H_0) were true. Essentially, it provides a standardized way to compare the observed outcomes against the expectations set by H_0, enabling us to assess whether the observed results are likely due to random chance or indicative of a significant effect.

The general formula for a test statistic is

$$\text{Test Statistic} = \frac{\text{Observed Value} - \text{Expected Value under } H_0}{\text{Standard Error}}$$

Each component of this formula has an important role:

1. **Numerator:**
   - The numerator represents the difference between the actual data (observed value) and the hypothetical value (expected value) that is assumed under H_0.
   - This difference quantifies the extent of the deviation. A larger deviation suggests stronger evidence against H_0.
2. **Denominator:**
   - The denominator is the *standard error*, which measures the variability or spread of the data. It accounts for factors such as sample size and the inherent randomness of the data.
   - By dividing the numerator by the standard error, the test statistic is standardized, allowing comparisons across different studies, sample sizes, and distributions.

The test statistic plays a central role in determining whether to reject H_0. Once calculated, it is compared to a known distribution (e.g., standard normal distribution for z-tests or t-distribution for t-tests). This comparison allows us to evaluate the likelihood of observing such a test statistic under H_0:

- **If the test statistic is close to 0:** This indicates that the observed data is very close to what is expected under H_0. There is little evidence to suggest rejecting H_0.
- **If the test statistic is far from 0 (in the tails of the distribution):** This suggests that the observed data deviates significantly from the expectations under H_0. Such deviations may provide strong evidence against H_0.

4.1.4.1 Why Standardizing Matters

Standardizing the difference between the observed and expected values ensures that the test statistic is not biased by factors such as the scale of measurement or the size of the sample. For instance:

- A raw difference of 5 might be highly significant in one context but negligible in another, depending on the variability (standard error).
- Standardizing ensures that the magnitude of the test statistic reflects both the size of the difference and the reliability of the sample data.

4.1.4.2 Interpreting the Test Statistic

After calculating the test statistic, it is used to:

1. Compare with a critical value: For example, in a z-test with $\alpha = 0.05$, the critical values are -1.96 and 1.96 for a two-tailed test. If the test statistic falls beyond these values, H_0 is rejected.
2. Calculate the p-value: The p-value is derived from the distribution and reflects the probability of observing a test statistic as extreme as the one calculated if H_0 is true.

4.1.5 Critical Values and Rejection Regions

The *critical value* is a point on the distribution that separates the rejection region from the non-rejection region:

- **Rejection Region**: If the test statistic falls in this region, we reject H_0.
- **Non-rejection Region**: If the test statistic falls here, we fail to reject H_0.

The rejection region depends on the significance level (α). For a two-tailed test with $\alpha = 0.05$, the critical values correspond to the top 2.5% and bottom 2.5% of the distribution.

4.1.6 Visualizing Hypothesis Testing

Let us create a visualization to tie these concepts together (Fig. 4.1):

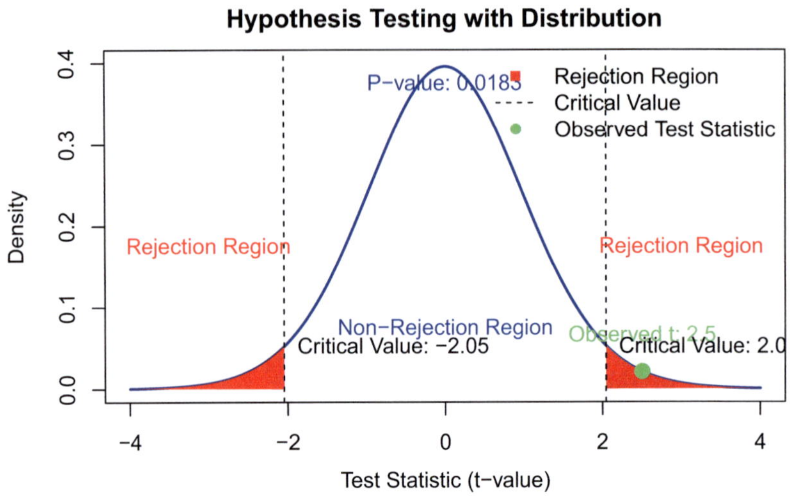

Fig. 4.1 Hypothesis testing with distribution

```
# Parameters
alpha <- 0.05   # Significance level
df <- 29        # Degrees of freedom (for t-distribution)
t_critical <-
    qt(1 - alpha / 2, df)  # Critical value for two-tailed
     ↪  test

# Generate t-distribution values
t_values <- seq(-4, 4, length.out = 1000)
density <- dt(t_values, df)

# Observed test statistic
t_obs <- 2.5  # Example observed test statistic

# Plot the t-distribution
plot(
    t_values,
    density,
    type = "l",
    lwd = 2,
    col = "blue",
    main = "Hypothesis Testing with Distribution",
    xlab = "Test Statistic (t-value)",
    ylab = "Density",
    ylim = c(0, 0.4)
```

```
)

# Shade the rejection regions
polygon(c(t_values[t_values <= -t_critical], -t_critical),
        c(density[t_values <= -t_critical], 0),
        col = "red",
        border = NA)
polygon(c(t_values[t_values >= t_critical], t_critical),
        c(density[t_values >= t_critical], 0),
        col = "red",
        border = NA)

# Add observed test statistic
points(
    t_obs,
    dt(t_obs, df),
    col = "green",
    pch = 19,
    cex = 1.5
)
text(
    t_obs,
    dt(t_obs, df) + 0.02,
    paste("Observed t:", round(t_obs, 2)),
    col = "green",
    pos = 3
)

# Highlight the critical values
abline(
    v = c(-t_critical, t_critical),
    col = "black",
    lty = 2
)
text(
    -t_critical,
    0.05,
    paste("Critical Value:", round(-t_critical, 2)),
    pos = 4,
    col = "black"
)
text(
    t_critical,
    0.05,
```

```
    paste("Critical Value:", round(t_critical, 2)),
    pos = 4,
    col = "black"
)

# Calculate p-value
p_value <- 2 * (1 - pt(abs(t_obs), df))  # Two-tailed p-value
text(0,
     0.35,
     paste("P-value:", round(p_value, 4)),
     col = "blue",
     pos = 3)

# Annotate regions
text(-3,
     0.15,
     "Rejection Region",
     col = "red",
     pos = 3)
text(3, 0.15, "Rejection Region", col = "red", pos = 3)
text(0,
     0.05,
     "Non-Rejection Region",
     col = "blue",
     pos = 3)

# Add legend
legend(
    "topright",
    legend = c("Rejection Region", "Critical Value", "Observed
    ↪  Test Statistic"),
    col = c("red", "black", "green"),
    lty = c(NA, 2, NA),
    pch = c(15, NA, 19),
    bty = "n"
)
```

4.2 Key Concepts and Definitions

4.2.1 Random Sample

A random sample of size n consists of n independent observations, each drawn from the same underlying population distribution. Independence ensures that no observation influences another, and identical distribution guarantees that all observations are governed by the same probability rules.

4.2.2 Sample Statistics

4.2.2.1 Sample Mean

The sample mean is a measure of central tendency:

$$\bar{X} = \frac{\sum_{i=1}^{n} X_i}{n}$$

- Example: Suppose we measure the heights of five individuals (in cm): 170, 165, 180, 175, 172. The sample mean is

$$\bar{X} = \frac{170 + 165 + 180 + 175 + 172}{5} = 172.4\,\text{cm}.$$

4.2.2.2 Sample Median

The sample median is the middle value of ordered data:

$$\tilde{x} = \begin{cases} \text{Middle observation,} & \text{if } n \text{ is odd,} \\ \text{Average of two middle observations,} & \text{if } n \text{ is even} \end{cases}$$

4.2.2.3 Sample Variance

The sample variance measures data spread:

$$S^2 = \frac{\sum_{i=1}^{n} (X_i - \bar{X})^2}{n - 1}$$

4.2.2.4 Sample Standard Deviation

The sample standard deviation is the square root of the variance:

$$S = \sqrt{S^2}$$

4.2.2.5 Sample Proportions

Used for categorical data:

$$\hat{p} = \frac{X}{n} = \frac{\text{Number of successes}}{\text{Sample size}}$$

4.2.2.6 Estimators

- **Point Estimator**: A statistic ($\hat{\theta}$) used to estimate a population parameter (θ)
- **Point Estimate**: The numerical value assumed by $\hat{\theta}$ when evaluated for a given sample
- **Unbiased Estimator**: A point estimator $\hat{\theta}$ is unbiased if $E(\hat{\theta}) = \theta$

Examples of unbiased estimators:

- $\bar{X}$ for μ (population mean)
- S^2 for σ^2 (population variance)
- $\hat{p}$ for p (population proportion)
- $\widehat{p_1 - p_2}$ for $p_1 - p_2$ (population proportion difference)
- $\bar{X}_1 - \bar{X}_2$ for $\mu_1 - \mu_2$ (population mean difference)

Note While S^2 is unbiased for σ^2, S is a biased estimator of σ.

4.2.3 *Distribution of the Sample Mean*

The sampling distribution of the mean $\bar{X}$ depends on:

1. **Population Distribution**:
   - If $X \sim N(\mu, \sigma^2)$, then $\bar{X} \sim N\left(\mu, \frac{\sigma^2}{n}\right)$.
2. **Central Limit Theorem**:
   - For large n, $\bar{X}$ approximately follows a normal distribution, regardless of the population's shape.

4.2.3.1 Standard Error of the Mean

The standard error quantifies variability in $\bar{X}$:

$$\sigma_{\bar{X}} = \frac{\sigma}{\sqrt{n}}$$

Example Suppose $\sigma = 10$ and $n = 25$. Then:

$$\sigma_{\bar{X}} = \frac{10}{\sqrt{25}} = 2.$$

The smaller the standard error, the more precise our estimate of the population mean.

4.3 One-Sample Inference

4.3.1 For Single Mean

Consider a scenario where

$$Y_i \sim \text{i.i.d. } N(\mu, \sigma^2),$$

where i.i.d. stands for "independent and identically distributed." This model can be expressed as

$$Y_i = \mu + \epsilon_i$$

where:

- $\epsilon_i \sim^{\text{i.i.d.}} N(0, \sigma^2)$.
- $E(Y_i) = \mu$.
- $\text{Var}(Y_i) = \sigma^2$.
- $\bar{y} \sim N(\mu, \sigma^2/n)$.

When σ^2 is estimated by s^2, the standardized test statistic follows a t-distribution:

$$\frac{\bar{y} - \mu}{s/\sqrt{n}} \sim t_{n-1}$$

A $100(1 - \alpha)\%$ confidence interval for μ is obtained as

$$1 - \alpha = P\left(-t_{\alpha/2;n-1} \leq \frac{\bar{y} - \mu}{s/\sqrt{n}} \leq t_{\alpha/2;n-1}\right),$$

Table 4.1 Confidence intervals, sample size, and test statistics for normal mean inference

Case	Confidence interval 100(1 – α)%	Sample size (confidence α, error d)	Hypothesis test statistic
σ^2 known, X normal (or $n \geq 25$)	$\bar{X} \pm z_{\alpha/2}\frac{\sigma}{\sqrt{n}}$	$n \approx \frac{z_{\alpha/2}^2\sigma^2}{d^2}$	$z = \frac{\bar{X}-\mu_0}{\sigma/\sqrt{n}}$
σ^2 unknown, X normal (or $n \geq 25$)	$\bar{X} \pm t_{\alpha/2}\frac{s}{\sqrt{n}}$	$n \approx \frac{z_{\alpha/2}^2 s^2}{d^2}$	$t = \frac{\bar{X}-\mu_0}{s/\sqrt{n}}$

or equivalently,

$$P\left(\bar{y} - t_{\alpha/2;n-1}\frac{s}{\sqrt{n}} \leq \mu \leq \bar{y} + t_{\alpha/2;n-1}\frac{s}{\sqrt{n}}\right).$$

The confidence interval is expressed as

$$\bar{y} \pm t_{\alpha/2;n-1}\frac{s}{\sqrt{n}}$$

where $s/\sqrt{n}$ is the standard error of $\bar{y}$.

If the experiment were repeated many times, 100(1 – α)% of these intervals would contain μ (Table 4.1).

4.3.1.1 Power in Hypothesis Testing

Power ($\pi(\mu)$) of a hypothesis test represents the probability of correctly rejecting the null hypothesis (H_0) when it is false (i.e., when alternative hypothesis H_A is true). Formally, it is expressed as

$$\begin{aligned} \text{Power} = \pi(\mu) &= 1 - \beta \\ &= P(\text{test rejects } H_0|\mu) \\ &= P(\text{test rejects } H_0|H_A \text{ is true}) \end{aligned}$$

where β is the probability of a type II error (failing to reject H_0 when it is false).

To calculate this probability:

1. **Under** H_0: The distribution of the test statistic is centered around the null parameter (e.g., μ_0).
2. **Under** H_A: The test statistic is distributed differently, shifted according to the true value under H_A (e.g., μ_1).

Hence, to evaluate the power, it is crucial to determine the distribution of the test statistic under the alternative hypothesis, H_A.

Below, we derive the power for both one-sided and two-sided z-tests.

4.3.1.1.1 One-Sided z-Test

Consider the hypotheses:

$$H_0 : \mu \leq \mu_0 \quad \text{vs.} \quad H_A : \mu > \mu_0$$

The power for a one-sided z-test is derived as follows:

1. The test rejects H_0 if $\bar{y} > \mu_0 + z_\alpha \frac{\sigma}{\sqrt{n}}$, where z_α is the critical value for the test at the significance level α.
2. Under the alternative hypothesis, the distribution of $\bar{y}$ is centered at μ, with standard deviation $\frac{\sigma}{\sqrt{n}}$.
3. The power is then

$$\begin{aligned}
\pi(\mu) &= P\left(\bar{y} > \mu_0 + z_\alpha \frac{\sigma}{\sqrt{n}} \middle| \mu\right) \\
&= P\left(Z > z_\alpha + \frac{\mu_0 - \mu}{\sigma/\sqrt{n}} \middle| \mu\right), \quad \text{where } Z = \frac{\bar{y} - \mu}{\sigma/\sqrt{n}} \\
&= 1 - \Phi\left(z_\alpha + \frac{(\mu_0 - \mu)\sqrt{n}}{\sigma}\right) \\
&= \Phi\left(-z_\alpha + \frac{(\mu - \mu_0)\sqrt{n}}{\sigma}\right).
\end{aligned}$$

Here, we use the symmetry of the standard normal distribution: $1 - \Phi(x) = \Phi(-x)$.

Suppose we wish to show that the mean response μ under the treatment is higher than the mean response μ_0 without treatment (i.e., the treatment effect $\delta = \mu - \mu_0$ is large).

Since power is an increasing function of $\mu - \mu_0$, it suffices to find the sample size n that achieves the desired power $1 - \beta$ at $\mu = \mu_0 + \delta$. The power at $\mu = \mu_0 + \delta$ is

$$\pi(\mu_0 + \delta) = \Phi\left(-z_\alpha + \frac{\delta\sqrt{n}}{\sigma}\right) = 1 - \beta$$

Given $\Phi(z_\beta) = 1 - \beta$, we have

$$-z_\alpha + \frac{\delta\sqrt{n}}{\sigma} = z_\beta$$

Solving for n, we obtain

$$n = \left(\frac{(z_\alpha + z_\beta)\sigma}{\delta}\right)^2$$

Larger sample sizes are required when:

- The sample variability is large (σ is large).
- The significance level α is small (z_α is large).
- The desired power $1 - \beta$ is large (z_β is large).
- The magnitude of the effect is small (δ is small).

In practice, δ and σ are often unknown. To estimate σ, you can:

1. Use prior studies or pilot studies.
2. Approximate σ based on the anticipated range of the observations (excluding outliers). For normally distributed data, dividing the range by four provides a reasonable estimate of σ.

These considerations ensure the test is adequately powered to detect meaningful effects, while balancing practical constraints such as sample size.

4.3.1.1.2 Two-Sided z-Test

For a two-sided test, the hypotheses are

$$H_0 : \mu = \mu_0 \quad \text{vs.} \quad H_A : \mu \neq \mu_0$$

The test rejects H_0 if $\bar{y}$ lies outside the interval $\mu_0 \pm z_{\alpha/2}\frac{\sigma}{\sqrt{n}}$. The power of the test is

$$\begin{aligned}\pi(\mu) &= P\left(\bar{y} < \mu_0 - z_{\alpha/2}\frac{\sigma}{\sqrt{n}}\middle|\mu\right) + P\left(\bar{y} > \mu_0 + z_{\alpha/2}\frac{\sigma}{\sqrt{n}}\middle|\mu\right)\\ &= \Phi\left(-z_{\alpha/2} + \frac{(\mu-\mu_0)\sqrt{n}}{\sigma}\right) + \Phi\left(-z_{\alpha/2} - \frac{(\mu-\mu_0)\sqrt{n}}{\sigma}\right)\end{aligned}$$

To ensure a power of $1-\beta$ when the treatment effect $\delta = |\mu-\mu_0|$ is at least a certain value, we solve for n. Since the power function for a two-sided test is increasing and symmetric in $|\mu - \mu_0|$, it suffices to find n such that the power equals $1 - \beta$ when $\mu = \mu_0 + \delta$. This gives

$$n = \left(\frac{(z_{\alpha/2} + z_\beta)\sigma}{\delta}\right)^2$$

Alternatively, the required sample size can be determined using a confidence interval approach. For a two-sided α-level confidence interval of the form:

$$\bar{y} \pm D$$

where $D = z_{\alpha/2}\frac{\sigma}{\sqrt{n}}$, solving for n gives

$$n = \left(\frac{z_{\alpha/2}\sigma}{D}\right)^2$$

This value should be rounded up to the nearest integer to ensure the required precision.

```
# Generate random data and compute a 95% confidence interval
data <- rnorm(100) # Generate 100 random values
t.test(data, conf.level = 0.95) # Perform t-test with 95%
↪  confidence interval
#>
#>  One Sample t-test
#>
#> data:  data
#> t = -1.2866, df = 99, p-value = 0.2012
#> alternative hypothesis: true mean is not equal to 0
#> 95 percent confidence interval:
#>  -0.34385554  0.07333768
#> sample estimates:
#>  mean of x
#> -0.1352589
```

For a one-sided hypothesis test, such as testing $H_0 : \mu \geq 30$ versus $H_a : \mu < 30$:

```
# Perform one-sided t-test
t.test(data, mu = 30, alternative = "less")
#>
#>  One Sample t-test
#>
#> data:  data
#> t = -286.65, df = 99, p-value < 2.2e-16
#> alternative hypothesis: true mean is less than 30
#> 95 percent confidence interval:
#>        -Inf 0.03929455
#> sample estimates:
#>  mean of x
#> -0.1352589
```

When σ is unknown, you can estimate it using:

1. Prior studies or pilot studies
2. The range of observations (excluding outliers) divided by 4, which provides a reasonable approximation for normally distributed data

4.3.1.1.3 z-Test Summary

- For one-sided tests:

$$\pi(\mu) = \Phi\left(-z_\alpha + \frac{(\mu - \mu_0)\sqrt{n}}{\sigma}\right)$$

- For two-sided tests:

$$\pi(\mu) = \Phi\left(-z_{\alpha/2} + \frac{(\mu - \mu_0)\sqrt{n}}{\sigma}\right) + \Phi\left(-z_{\alpha/2} - \frac{(\mu - \mu_0)\sqrt{n}}{\sigma}\right)$$

Factors Affecting Power:

- **Effect Size** $(\mu - \mu_0)$: Larger differences between μ and μ_0 increase power.
- **Sample Size** (n): Larger n reduces the standard error, increasing power.
- **Variance** (σ^2): Smaller variance increases power.
- **Significance Level** (α): Increasing α (making the test more liberal) increases power through z_α.

4.3.1.1.4 One-Sample t-Test

In hypothesis testing, calculating the power and determining the required sample size for *t-tests* are more complex than those for *z-tests*. This complexity arises from the involvement of the *Student's t-distribution* and its generalized form, the *noncentral t-distribution*.

The power function for a one-sample t-test can be expressed as

$$\pi(\mu) = P\left(\frac{\bar{y} - \mu_0}{s/\sqrt{n}} > t_{n-1;\alpha} \mid \mu\right)$$

Here:

- μ_0 is the hypothesized population mean under the null hypothesis.
- $\bar{y}$ is the sample mean.
- s is the sample standard deviation.
- n is the sample size.
- $t_{n-1;\alpha}$ is the critical t-value from the Student's t-distribution with $n - 1$ degrees of freedom at significance level α.

When $\mu > \mu_0$ (i.e., $\mu - \mu_0 = \delta$), the random variable

$$T = \frac{\bar{y} - \mu_0}{s/\sqrt{n}}$$

does not follow the Student's t-distribution. Instead, it follows a *noncentral t-distribution* with:

- A *non-centrality parameter* $\lambda = \delta\sqrt{n}/\sigma$, where σ is the population standard deviation
- Degrees of freedom $n - 1$

Key Properties of the Power Function:

- The power $\pi(\mu)$ is an increasing function of the non-centrality parameter λ.
- For $\delta = 0$ (i.e., when the null hypothesis is true), the noncentral t-distribution simplifies to the regular Student's t-distribution.

To calculate the power in practice, numerical procedures (see below) or precomputed charts are typically required.

Approximate Sample Size Adjustment for t-Tests

When planning a study, researchers often start with an approximation based on *z-tests* and then adjust for the specifics of the t-test. Here is the process:

1. Start with the sample size for a z-test.
   For a two-sided test:

$$n_z = \frac{\left(z_{\alpha/2} + z_\beta\right)^2 \sigma^2}{\delta^2}$$

   where:

   - $z_{\alpha/2}$ is the critical value from the standard normal distribution for a two-tailed test.
   - z_β corresponds to the desired power $1 - \beta$.
   - δ is the effect size $\mu - \mu_0$.
   - σ is the population standard deviation.

2. Adjust for the t-distribution.
   Let $v = n - 1$, where n is the sample size derived from the z-test. For a two-sided t-test, the approximate sample size is

$$n^* = \frac{\left(t_{v;\alpha/2} + t_{v;\beta}\right)^2 \sigma^2}{\delta^2}$$

   Here:

- $t_{v;\alpha/2}$ and $t_{v;\beta}$ are the critical values from the Student's t-distribution for the significance level α and desired power, respectively.
- Since v depends on n^*, this process may require iterative refinement.

Notes:

1. **Approximations**: The above formulas provide an intuitive starting point, but may require adjustments based on exact numerical solutions.
2. **Insights**: Power is an increasing function of:
   - The effect size δ
   - The sample size n
   - and A decreasing function of the population variability σ

```
# Example: Power calculation for a one-sample t-test
library(pwr)

# Parameters
effect_size <- 0.5  # Cohen's d
alpha <- 0.05       # Significance level
power <- 0.8        # Desired power

# Compute sample size
sample_size <-
    pwr.t.test(
        d = effect_size,
        sig.level = alpha,
        power = power,
        type = "one.sample"
    )$n

# Print result
cat("Required sample size for one-sample t-test:",
    ceiling(sample_size),
    "\n")
#> Required sample size for one-sample t-test: 34

# Power calculation for a given sample size
calculated_power <-
    pwr.t.test(
        n = ceiling(sample_size),
        d = effect_size,
        sig.level = alpha,
        type = "one.sample"
    )$power
cat("Achieved power with computed sample size:",
    calculated_power,
    "\n")
#> Achieved power with computed sample size: 0.8077775
```

Table 4.2 Inference for the difference between two means under various variance assumptions

	100(1 − α) confidence interval	Hypothesis testing test statistic	
When σ^2 is known	$\bar{X}_1 - \bar{X}_2 \pm z_{\alpha/2}\sqrt{\frac{\sigma_1^2}{n_1} + \frac{\sigma_2^2}{n_2}}$	$z = \frac{(\bar{X}_1-\bar{X}_2)-(\mu_1-\mu_2)_0}{\sqrt{\frac{\sigma_1^2}{n_1}+\frac{\sigma_2^2}{n_2}}}$	
When σ^2 is unknown, variances assumed EQUAL	$\bar{X}_1 - \bar{X}_2 \pm t_{\alpha/2}\sqrt{s_p^2(\frac{1}{n_1} + \frac{1}{n_2})}$	$t = \frac{(\bar{X}_1-\bar{X}_2)-(\mu_1-\mu_2)_0}{\sqrt{s_p^2(\frac{1}{n_1}+\frac{1}{n_2})}}$	Pooled variance: $s_p^2 = \frac{(n_1-1)s_1^2-(n_2-1)s_2^2}{n_1+n_2-2}$ Degrees of freedom: $\gamma = n_1 + n_2 - 2$
When σ^2 is unknown, variances assumed UNEQUAL	$\bar{X}_1 - \bar{X}_2 \pm t_{\alpha/2}\sqrt{(\frac{s_1^2}{n_1} + \frac{s_2^2}{n_2})}$	$t = \frac{(\bar{X}_1-\bar{X}_2)-(\mu_1-\mu_2)_0}{\sqrt{(\frac{s_1^2}{n_1}+\frac{s_2^2}{n_2})}}$	Degrees of freedom: $\gamma = \frac{(\frac{s_1^2}{n_1}+\frac{s_2^2}{n_2})^2}{\frac{(\frac{s_1^2}{n_1})^2}{n_1-1}+\frac{(\frac{s_2^2}{n_2})^2}{n_2-1}}$

Table 4.3 Confidence interval and test statistic for paired sample mean difference

Metric	Formula
Confidence interval	$\bar{D} \pm t_{\alpha/2}\frac{s_d}{\sqrt{n}}$
Hypothesis test statistic	$t = \frac{\bar{D}-D_0}{s_d/\sqrt{n}}$

4.3.2 *For Difference of Means, Independent Samples*

See Table 4.2.

4.3.3 *For Difference of Means, Paired Samples*

See Table 4.3.

4.3.4 *For Difference of Two Proportions*

The mean of the difference between two sample proportions is given by

$$\hat{p}_1 - \hat{p}_2$$

The variance of the difference in proportions is

$$\frac{p_1(1-p_1)}{n_1} + \frac{p_2(1-p_2)}{n_2}$$

A $100(1-\alpha)\%$ confidence interval for the difference in proportions is calculated as

$$\hat{p}_1 - \hat{p}_2 \pm z_{\alpha/2}\sqrt{\frac{p_1(1-p_1)}{n_1} + \frac{p_2(1-p_2)}{n_2}}$$

where:

- $z_{\alpha/2}$: the critical value from the standard normal distribution
- $\hat{p}_1, \hat{p}_2$: sample proportions
- n_1, n_2: sample sizes

Sample Size for a Desired Confidence Level and Margin of Error
To achieve a margin of error d for a given confidence level, the required sample size can be estimated as follows:

1. **With Prior Estimates of** $\hat{p}_1$ and $\hat{p}_2$:

$$n \approx \frac{z_{\alpha/2}^2\,[p_1(1-p_1) + p_2(1-p_2)]}{d^2}.$$

2. **Without Prior Estimates** (assuming maximum variability, $\hat{p} = 0.5$):

$$n \approx \frac{z_{\alpha/2}^2}{2d^2}.$$

Hypothesis Testing for Difference in Proportions
The test statistic for hypothesis testing depends on the null hypothesis:

1. *When* $(p_1 - p_2) \neq 0$:

$$z = \frac{(\hat{p}_1 - \hat{p}_2) - (p_1 - p_2)_0}{\sqrt{\frac{p_1(1-p_1)}{n_1} + \frac{p_2(1-p_2)}{n_2}}}.$$

2. *When* $(p_1 - p_2)_0 = 0$ (testing equality of proportions):

$$z = \frac{\hat{p}_1 - \hat{p}_2}{\sqrt{\hat{p}(1-\hat{p})\left(\frac{1}{n_1} + \frac{1}{n_2}\right)}}$$

where $\hat{p}$ is the pooled sample proportion:

$$\hat{p} = \frac{x_1 + x_2}{n_1 + n_2} = \frac{n_1\hat{p}_1 + n_2\hat{p}_2}{n_1 + n_2}.$$

4.3.5 For Single Proportion

The $100(1-\alpha)\%$ confidence interval for a population proportion p is

$$\hat{p} \pm z_{\alpha/2}\sqrt{\frac{\hat{p}(1-\hat{p})}{n}}$$

Sample Size Determination:

- **With Prior Estimate** ($\hat{p}$):

$$n \approx \frac{z_{\alpha/2}^2 \hat{p}(1-\hat{p})}{d^2}.$$

- **Without Prior Estimate**:

$$n \approx \frac{z_{\alpha/2}^2}{4d^2}.$$

The test statistic for $H_0 : p = p_0$ is

$$z = \frac{\hat{p} - p_0}{\sqrt{\frac{p_0(1-p_0)}{n}}}$$

4.3.6 For Single Variance

For a sample variance s^2 with n observations, the $100(1-\alpha)\%$ confidence interval for the population variance σ^2 is

$$\begin{aligned} 1-\alpha &= P(\chi_{1-\alpha/2;n-1}^2) \leq (n-1)s^2/\sigma^2 \leq \chi_{\alpha/2;n-1}^2) \\ &= P\left(\frac{(n-1)s^2}{\chi_{\alpha/2;n-1}^2} \leq \sigma^2 \leq \frac{(n-1)s^2}{\chi_{1-\alpha/2;n-1}^2}\right) \end{aligned}$$

Equivalently, the confidence interval can be written as

$$\left(\frac{(n-1)s^2}{\chi_{\alpha/2}^2}, \frac{(n-1)s^2}{\chi_{1-\alpha/2}^2}\right)$$

To find confidence limits for σ, compute the square root of the interval bounds:

$$\text{Confidence Interval for } \sigma: \quad \left(\sqrt{\frac{(n-1)s^2}{\chi^2_{\alpha/2}}}, \sqrt{\frac{(n-1)s^2}{\chi^2_{1-\alpha/2}}}\right)$$

Hypothesis Testing for Variance
The test statistic for testing a null hypothesis about a population variance (σ_0^2) is

$$\chi^2 = \frac{(n-1)s^2}{\sigma_0^2}$$

This test statistic follows a chi-squared distribution with $n-1$ degrees of freedom under the null hypothesis.

4.3.7 *Nonparametric Tests*

See Table 4.4.

4.3.7.1 Sign Test

The *sign test* is used to test hypotheses about the median of a population, $\mu_{(0.5)}$, without assuming a specific distribution for the data. This test is ideal for small sample sizes or when normality assumptions are not met.

To test the population median, consider the hypotheses:

- Null Hypothesis: $H_0 : \mu_{(0.5)} = 0$.
- Alternative Hypothesis: $H_a : \mu_{(0.5)} > 0$ (one-sided test).

Table 4.4 Nonparametric tests for medians, quantiles, and randomness with assumptions

Method	Purpose	Assumptions
Sign test	Test median	None (ordinal data sufficient)
Wilcoxon signed rank test	Test symmetry around a value	Symmetry of distribution
Wald-Wolfowitz Runs Test	Test for randomness	Independent observations
Quantile (or percentile) test	Test specific quantile	None (ordinal data sufficient)

Steps:

1. **Count Positive and Negative Deviations**:
   - Count observations (y_i) greater than 0: s_+ (number of positive signs).
   - Count observations less than 0: s_- (number of negative signs).
   - $s_- = n - s_+$.
2. **Decision Rule**:
   - Reject H_0 if s_+ is large (or equivalently, s_- is small).
   - To determine how large s_+ must be, use the distribution of S_+ under H_0, which is *binomial* with $p = 0.5$.
3. **Null Distribution**:
   Under H_0, S_+ follows:
$$S_+ \sim Binomial(n, p = 0.5).$$
4. **Critical Value**:
   Reject H_0 if:
$$s_+ \geq b_{n,\alpha}$$
   where $b_{n,\alpha}$ is the upper α critical value of the binomial distribution.
5. **p-Value Calculation**:
   Compute the p-value for the observed (one-tailed) s_+ as:
$$\text{p-value} = P(S \geq s_+) = \sum_{i=s_+}^{n} \binom{n}{i} \left(\frac{1}{2}\right)^n$$
   Alternatively:
$$P(S \leq s_-) = \sum_{i=0}^{s_-} \binom{n}{i} \left(\frac{1}{2}\right)^n$$

Large Sample Normal Approximation:
For large n, use a normal approximation for the binomial test. Reject H_0 if:

$$s_+ \geq \frac{n}{2} + \frac{1}{2} + z_\alpha \sqrt{\frac{n}{4}}$$

where z_α is the critical value for a one-sided test.
For two-sided tests, use the maximum or minimum of s_+ and s_-:

- Test statistic: $s_{\max} = \max(s_+, s_-)$ or $s_{\min} = \min(s_+, s_-)$.

- Reject H_0 if p-value is less than α, where:

$$p\text{-value} = 2\sum_{i=s_{\max}}^{n}\binom{n}{i}\left(\frac{1}{2}\right)^n = 2\sum_{i=0}^{s_{\min}}\binom{n}{i}\left(\frac{1}{2}\right)^n.$$

Equivalently, rejecting H_0 if $s_{\max} \geq b_{n,\alpha/2}$.
For large n, the normal approximation uses:

$$z = \frac{s_{\max} - \frac{n}{2} - \frac{1}{2}}{\sqrt{\frac{n}{4}}}$$

Reject H_0 at α if $z \geq z_{\alpha/2}$.
Handling zeros in the data is a common issue with the sign test:

1. **Random Assignment**: Assign zeros randomly to either s_+ or s_- (two researchers might get different results).
2. **Fractional Assignment**: Count each zero as 0.5 toward both s_+ and s_- (but then we could not apply the binomial distribution afterward).
3. **Ignore Zeros**: Ignore zeros, but note this reduces the sample size and power.

```
# Example Data
data <- c(0.76, 0.82, 0.80, 0.79, 1.06, 0.83, -0.43, -0.34,
↪  3.34, 2.33)

# Count positive signs
s_plus <- sum(data > 0)

# Sample size excluding zeros
n <- length(data)

# Perform a one-sided binomial test
binom.test(s_plus, n, p = 0.5, alternative = "greater")
#>
#>  Exact binomial test
#>
#> data:  s_plus and n
#> number of successes = 8, number of trials = 10,
#> p-value = 0.05469
#> alternative hypothesis: true probability of success is
↪  greater than 0.5
#> 95 percent confidence interval:
#>  0.4930987 1.0000000
#> sample estimates:
#> probability of success
#>                    0.8
```

4.3.7.2 Wilcoxon Signed Rank Test

The *Wilcoxon signed rank test* is an improvement over the sign test as it considers both the magnitude and direction of deviations from the null hypothesis value (e.g., 0). However, this test assumes that the data are symmetrically distributed around the median, unlike the sign test.

We test the following hypotheses:

$$H_0 : \mu_{(0.5)} = 0$$
$$H_a : \mu_{(0.5)} > 0$$

This example assumes no ties or duplicate observations in the data.
Procedure for the signed rank test

1. **Rank the Absolute Values**:
   - Rank the observations y_i based on their absolute values.
   - Let r_i denote the rank of y_i.
   - Since there are no ties, ranks r_i are uniquely determined and form a permutation of integers $1, 2, \ldots, n$.
2. *Calculate* w_+ and w_-:
   - w_+ is the sum of the ranks corresponding to positive values of y_i.
   - w_- is the sum of the ranks corresponding to negative values of y_i.
   - By definition:

$$w_+ + w_- = \sum_{i=1}^{n} r_i = \frac{n(n+1)}{2}.$$

3. **Decision Rule**:
   - Reject H_0 if w_+ is large (or equivalently, if w_- is small).

Null Distribution of W_+:
Under the null hypothesis, the distributions of W_+ and W_- are identical and symmetric. The p-value for a one-sided test is

$$\text{p-value} = P(W \geq w_+) = P(W \leq w_-)$$

An α-level test rejects H_0 if $w_+ \geq w_{n,\alpha}$, where $w_{n,\alpha}$ is the critical value from a table of the null distribution of W_+.

For two-sided tests, use

$$p\text{-value} = 2P(W \geq w_{\max}) = 2P(W \leq w_{\min})$$

Normal Approximation for Large Samples:
For large n, the null distribution of W_+ can be approximated by a normal distribution:

$$z = \frac{w_+ - \frac{n(n+1)}{4} - \frac{1}{2}}{\sqrt{\frac{n(n+1)(2n+1)}{24}}}$$

The test rejects H_0 at level α if

$$w_+ \geq \frac{n(n+1)}{4} + \frac{1}{2} + z_\alpha \sqrt{\frac{n(n+1)(2n+1)}{24}} \approx w_{n,\alpha}$$

For a two-sided test, the decision rule uses the maximum or minimum of w_+ and w_-:

- $w_{\max} = \max(w_+, w_-)$.
- $w_{\min} = \min(w_+, w_-)$.

The p-value is computed as

$$p\text{-value} = 2P(W \geq w_{\max}) = 2P(W \leq w_{\min})$$

Handling Tied Ranks:
If some observations $|y_i|$ have tied absolute values, assign the average rank (or "midrank") to all tied values. For example:

- Suppose $y_1 = -1$, $y_2 = 3$, $y_3 = -3$, and $y_4 = 5$.
- The ranks for $|y_i|$ are:
  - $|y_1| = 1$: $r_1 = 1$.
  - $|y_2| = |y_3| = 3$: $r_2 = r_3 = \frac{2+3}{2} = 2.5$.
  - $|y_4| = 5$: $r_4 = 4$.

```
# Example Data
data <- c(0.76, 0.82, 0.80, 0.79, 1.06, 0.83, -0.43, -0.34,
↪   3.34, 2.33)

# Perform Wilcoxon Signed Rank Test (exact test)
wilcox_exact <- wilcox.test(data, exact = TRUE)

# Display results
wilcox_exact
#>
#>  Wilcoxon signed rank exact test
#>
```

```
#> data:  data
#> V = 52, p-value = 0.009766
#> alternative hypothesis: true location is not equal to 0
```

For large samples, you can use the normal approximation by setting `exact = FALSE`:

```
# Perform Wilcoxon Signed Rank Test (normal approximation)
wilcox_normal <- wilcox.test(data, exact = FALSE)

# Display results
wilcox_normal
#>
#>  Wilcoxon signed rank test with continuity correction
#>
#> data:  data
#> V = 52, p-value = 0.01443
#> alternative hypothesis: true location is not equal to 0
```

4.3.7.3 Wald-Wolfowitz Runs Test

The *runs test* is a nonparametric test used to examine the randomness of a sequence. Specifically, it tests whether the order of observations in a sequence is random. This test is useful in detecting nonrandom patterns, such as trends, clustering, or periodicity.
The hypotheses for the runs test are:

- Null Hypothesis: H_0: The sequence is random.
- Alternative Hypothesis: H_a: The sequence is not random.

A *run* is a sequence of consecutive observations of the same type. For example: In the binary sequence + + - - + - + +, there are *five runs*: ++, --, +, -, ++.
Runs can be formed based on any classification criteria, such as:

- Positive vs. negative values
- Above vs. below the median
- Success vs. failure in binary outcomes

Test Statistic:

1. *Number of Runs* (R):
   The observed number of runs in the sequence
2. *Expected Number of Runs* ($E[R]$):

Under the null hypothesis of randomness, the expected number of runs is

$$E[R] = \frac{2n_1n_2}{n_1 + n_2} + 1$$

where:

- n_1: number of observations in the first category (e.g., positives)
- n_2: number of observations in the second category (e.g., negatives)
- $n = n_1 + n_2$: total number of observations

3. **Variance of Runs** (Var[R]):
   The variance of the number of runs is given by:

$$\text{Var}[R] = \frac{2n_1n_2(2n_1n_2 - n)}{n^2(n - 1)}$$

4. **Standardized Test Statistic** (z):
   For large samples ($n \geq 20$), the test statistic is approximately normally distributed:

$$z = \frac{R - E[R]}{\sqrt{\text{Var}[R]}}$$

Decision Rule:

- Compute the z-value and compare it to the critical value of the standard normal distribution.
- For a significance level α:
  - Reject H_0 if $|z| \geq z_{\alpha/2}$ (two-sided test).
  - Reject H_0 if $z \geq z_\alpha$ or $z \leq -z_\alpha$ for one-sided tests.

Steps for Conducting a Runs Test:

1. Classify the data into two groups (e.g., above/below median, positive/negative).
2. Count the total number of runs (R).
3. Compute $E[R]$ and Var[R] based on n_1 and n_2.
4. Compute the z-value for the observed number of runs.
5. Compare the z-value to the critical value to decide whether to reject H_0.

For a numerical dataset where the test is based on values above and below the median:

```r
# Example dataset
data <- c(1.2, -0.5, 3.4, -1.1, 2.8, -0.8, 4.5, 0.7)

library(randtests)
# Perform Runs Test (above/below median)
runs.test(data)
#>
#>  Runs Test
#>
#> data:  data
#> statistic = 2.2913, runs = 8, n1 = 4, n2 = 4, n = 8,
#> p-value = 0.02195
#> alternative hypothesis: nonrandomness
```

The output of the `runs.test` function includes:

- **Observed Runs**: The actual number of runs in the sequence.
- **Expected Runs**: The expected number of runs under H_0.
- **p-value**: The probability of observing a number of runs as extreme as the observed one under H_0.
- If the p-value is less than α, reject H_0 and conclude that the sequence is not random.

Limitations of the Runs Test:

- The test assumes that observations are independent.
- For small sample sizes, the test may have limited power.
- Ties in the data must be resolved by a predefined rule (e.g., treating ties as belonging to one group or excluding them).

4.3.7.4 Quantile (or Percentile) Test

The *quantile test* (also called the percentile test) is a nonparametric test used to evaluate whether the proportion of observations falling within a specific quantile matches the expected proportion under the null hypothesis. This test is useful for assessing the distribution of data when specific quantiles (e.g., medians or percentiles) are of interest.

Suppose we want to test whether the true proportion of data below a specified quantile q matches a given probability p. The hypotheses are:

- Null Hypothesis: H_0: The true proportion is equal to p.
- Alternative Hypothesis: H_a: The true proportion is not equal to p (two-sided), greater than p (right-tailed), or less than p (left-tailed).

Test Statistic

The test statistic is based on the observed count of data points below the specified quantile:

1. **Observed Count** (k):
   The number of data points y_i such that $y_i \leq q$.
2. **Expected Count** ($E[k]$):
   The expected number of observations below the quantile q under H_0 is

$$E[k] = n \cdot p$$

3. **Variance**:
   Under the binomial distribution, the variance is

$$\text{Var}[k] = n \cdot p \cdot (1 - p)$$

4. **Standardized Test Statistic** (z):
   For large n, the test statistic is approximately normally distributed:

$$z = \frac{k - E[k]}{\sqrt{\text{Var}[k]}} = \frac{k - n \cdot p}{\sqrt{n \cdot p \cdot (1 - p)}}$$

Decision Rule:

1. Compute the z-value for the observed count.
2. Compare the z-value to the critical value of the standard normal distribution:
   - For a two-sided test, reject H_0 if $|z| \geq z_{\alpha/2}$.
   - For a one-sided test, reject H_0 if $z \geq z_\alpha$ (right-tailed) or $z \leq -z_\alpha$ (left-tailed).

Alternatively, calculate the p-value and reject H_0 if the p-value $\leq \alpha$.

Suppose we have a dataset and want to test whether the proportion of observations below the 50th percentile (median) matches the expected value of $p = 0.5$.

```
# Example data
data <- c(12, 15, 14, 10, 13, 11, 14, 16, 15, 13)

# Define the quantile to test
quantile_value <- quantile(data, 0.5) # Median
p <- 0.5                              # Proportion under H0

# Count observed values below or equal to the quantile
k <- sum(data <= quantile_value)

# Sample size
n <- length(data)
```

```
# Expected count under H0
expected_count <- n * p

# Variance
variance <- n * p * (1 - p)

# Test statistic (z-value)
z <- (k - expected_count) / sqrt(variance)

# Calculate p-value for two-sided test
p_value <- 2 * (1 - pnorm(abs(z)))

# Output results
list(
  quantile_value = quantile_value,
  observed_count = k,
  expected_count = expected_count,
  z_value = z,
  p_value = p_value
)
#> $quantile_value
#>  50%
#> 13.5
#>
#> $observed_count
#> [1] 5
#>
#> $expected_count
#> [1] 5
#>
#> $z_value
#> [1] 0
#>
#> $p_value
#> [1] 1
```

For a one-sided test (e.g., testing whether the proportion is greater than p):

```
# Calculate one-sided p-value
p_value_one_sided <- 1 - pnorm(z)

# Output one-sided p-value
p_value_one_sided
#> [1] 0.5
```

Interpretation of Results:

- **p-value**: If the p-value is less than α, reject H_0 and conclude that the proportion of observations below the quantile deviates significantly from p.
- **Quantile Test Statistic** (z): The z-value indicates how many standard deviations the observed count is from the expected count under the null hypothesis. Large positive or negative z-values suggest nonrandom deviations.

Assumptions of the Test:

1. Observations are independent.
2. The sample size is large enough for the normal approximation to the binomial distribution to be valid ($n \cdot p \geq 5$ and $n \cdot (1 - p) \geq 5$).

Limitations of the Test:

- For small sample sizes, the normal approximation may not hold. In such cases, exact binomial tests are more appropriate.
- The test assumes that the quantile used (e.g., the median) is well-defined and correctly calculated from the data.

4.4 Two-Sample Inference

4.4.1 For Means

Suppose we have two sets of observations:

- $y_1, \ldots, y_{n_y}$
- $x_1, \ldots, x_{n_x}$

These are random samples from two independent populations with means μ_y and μ_x and variances σ_y^2 and σ_x^2. Our goal is to compare μ_y and μ_x or test whether $\sigma_y^2 = \sigma_x^2$.

4.4.1.1 Large Sample Tests

If n_y and n_x are large ($\geq$30), the central limit theorem allows us to make the following assumptions:

- **Expectation**:

$$E(\bar{y} - \bar{x}) = \mu_y - \mu_x.$$

- **Variance**:

$$\mathrm{Var}(\bar{y} - \bar{x}) = \frac{\sigma_y^2}{n_y} + \frac{\sigma_x^2}{n_x}.$$

The test statistic is

$$Z = \frac{\bar{y} - \bar{x} - (\mu_y - \mu_x)}{\sqrt{\frac{\sigma_y^2}{n_y} + \frac{\sigma_x^2}{n_x}}} \sim N(0, 1).$$

For large samples, replace variances with their unbiased estimators s_y^2 and s_x^2, yielding the same large sample distribution.

Confidence Interval
An approximate $100(1-\alpha)\%$ confidence interval for $\mu_y - \mu_x$ is

$$\bar{y} - \bar{x} \pm z_{\alpha/2}\sqrt{\frac{s_y^2}{n_y} + \frac{s_x^2}{n_x}}$$

Hypothesis Test
Testing:

$$H_0 : \mu_y - \mu_x = \delta_0 \quad \text{vs.} \quad H_a : \mu_y - \mu_x \neq \delta_0$$

The test statistic:

$$z = \frac{\bar{y} - \bar{x} - \delta_0}{\sqrt{\frac{s_y^2}{n_y} + \frac{s_x^2}{n_x}}}$$

Reject H_0 at the α level if:

$$|z| > z_{\alpha/2}$$

If $\delta_0 = 0$, this tests whether the two means are equal.

```
# Large sample test
y <- c(10, 12, 14, 16, 18)
x <- c(9, 11, 13, 15, 17)

# Mean and variance
mean_y <- mean(y)
```

```r
mean_x <- mean(x)
var_y <- var(y)
var_x <- var(x)
n_y <- length(y)
n_x <- length(x)

# Test statistic
z <- (mean_y - mean_x) / sqrt(var_y / n_y + var_x / n_x)
p_value <- 2 * (1 - pnorm(abs(z)))

list(z = z, p_value = p_value)
#> $z
#> [1] 0.5
#>
#> $p_value
#> [1] 0.6170751
```

4.4.1.2 Small Sample Tests

If the samples are small, assume the data come from independent normal distributions:

- $y_i \sim N(\mu_y, \sigma_y^2)$
- $x_i \sim N(\mu_x, \sigma_x^2)$

We can do inference based on the Student's T-distribution, where we have two cases:

- Equal variances
- Unequal variances (Table 4.5)

4.4.1.2.1 Equal Variances

Assumptions:

1. Independence and Identically Distributed (i.i.d.) Observations

Assume that observations in each sample are i.i.d., which implies

$$var(\bar{y}) = \frac{\sigma_y^2}{n_y}, \quad var(\bar{x}) = \frac{\sigma_x^2}{n_x}$$

2. Independence Between Samples

Table 4.5 Statistical assumptions, diagnostic tests, and graphical methods

Assumption	Tests	Plots
Independence and identically distributed (i.i.d.) observations	Test for serial correlation	
Independence between samples	Correlation coefficient	Scatterplot
Normality	See Normality Assessment	See Normality Assessment
Equality of variances	1. F-test 2. Levene's test 3. Modified Levene's test (Brown-Forsythe test) 4. Bartlett's test	1. Boxplots with overlayed means 2. Residuals spread plots

The samples are assumed to be independent, meaning no observation from one sample influences observations from the other. This independence allows us to write:

$$\begin{aligned} var(\bar{y} - \bar{x}) &= var(\bar{y}) + var(\bar{x}) - 2cov(\bar{y}, \bar{x}) \\ &= var(\bar{y}) + var(\bar{x}) \\ &= \frac{\sigma_y^2}{n_y} + \frac{\sigma_x^2}{n_x} \end{aligned}$$

This calculation assumes $cov(\bar{y}, \bar{x}) = 0$ due to the independence between the samples:

3. Normality Assumption

We assume that the underlying populations are normally distributed. This assumption justifies the use of the Student's T-distribution, which is critical for hypothesis testing and constructing confidence intervals.

4. Equality of Variances

If the population variances are equal, i.e., $\sigma_y^2 = \sigma_x^2 = \sigma^2$, then s_y^2 and s_x^2 are both unbiased estimators of σ^2. This allows us to pool the variances.
The pooled variance estimator is calculated as

$$s^2 = \frac{(n_y - 1)s_y^2 + (n_x - 1)s_x^2}{(n_y - 1) + (n_x - 1)}$$

The pooled variance estimate has degrees of freedom equal to

$$df = (n_y + n_x - 2)$$

Test Statistic
The test statistic is

$$T = \frac{\bar{y} - \bar{x} - (\mu_y - \mu_x)}{s\sqrt{\frac{1}{n_y} + \frac{1}{n_x}}} \sim t_{n_y+n_x-2}$$

Confidence Interval
A $100(1-\alpha)\%$ confidence interval for $\mu_y - \mu_x$ is:

$$\bar{y} - \bar{x} \pm t_{n_y+n_x-2,\alpha/2} \cdot s\sqrt{\frac{1}{n_y} + \frac{1}{n_x}}$$

Hypothesis Test
Testing:

$$H_0 : \mu_y - \mu_x = \delta_0 \quad \text{vs.} \quad H_a : \mu_y - \mu_x \neq \delta_0$$

Reject H_0 if:

$$|T| > t_{n_y+n_x-2,\alpha/2}$$

```
# Small sample test with equal variance
t_test_equal <- t.test(y, x, var.equal = TRUE)
t_test_equal
#>
#>  Two Sample t-test
#>
#> data:  y and x
#> t = 0.5, df = 8, p-value = 0.6305
#> alternative hypothesis: true difference in means is not
↪  equal to 0
#> 95 percent confidence interval:
#>  -3.612008  5.612008
#> sample estimates:
#> mean of x mean of y
#>        14        13
```

4.4.1.2.2 Unequal Variances

Assumptions:

1. Independence and Identically Distributed (i.i.d.) Observations

Assume that observations in each sample are i.i.d., which implies

$$var(\bar{y}) = \frac{\sigma_y^2}{n_y}, \quad var(\bar{x}) = \frac{\sigma_x^2}{n_x}$$

2. Independence Between Samples

The samples are assumed to be independent, meaning no observation from one sample influences observations from the other. This independence allows us to write:

$$\begin{aligned} var(\bar{y} - \bar{x}) &= var(\bar{y}) + var(\bar{x}) - 2cov(\bar{y}, \bar{x}) \\ &= var(\bar{y}) + var(\bar{x}) \\ &= \frac{\sigma_y^2}{n_y} + \frac{\sigma_x^2}{n_x} \end{aligned}$$

This calculation assumes $cov(\bar{y}, \bar{x}) = 0$ due to the independence between the samples.

3. Normality Assumption

We assume that the underlying populations are normally distributed. This assumption justifies the use of the Student's T-distribution, which is critical for hypothesis testing and constructing confidence intervals.

4. Unequal Variances

$\sigma_y^2 \neq \sigma_x^2$

Test Statistic
The test statistic is

$$T = \frac{\bar{y} - \bar{x} - (\mu_y - \mu_x)}{\sqrt{\frac{s_y^2}{n_y} + \frac{s_x^2}{n_x}}}$$

Degrees of Freedom (Welch-Satterthwaite Approximation) (Satterthwaite, 1946)
The degrees of freedom are approximated by

$$v = \frac{\left(\frac{s_y^2}{n_y} + \frac{s_x^2}{n_x}\right)^2}{\frac{\left(\frac{s_y^2}{n_y}\right)^2}{n_y - 1} + \frac{\left(\frac{s_x^2}{n_x}\right)^2}{n_x - 1}}$$

Since v is fractional, truncate to the nearest integer.

Confidence Interval
A $100(1-\alpha)\%$ confidence interval for $\mu_y - \mu_x$ is

$$\bar{y} - \bar{x} \pm t_{v,\alpha/2}\sqrt{\frac{s_y^2}{n_y} + \frac{s_x^2}{n_x}}$$

Hypothesis Test
Testing:

$$H_0 : \mu_y - \mu_x = \delta_0 \quad \text{vs.} \quad H_a : \mu_y - \mu_x \neq \delta_0$$

Reject H_0 if:

$$|T| > t_{v,\alpha/2}$$

where

$$t = \frac{\bar{y} - \bar{x} - \delta_0}{\sqrt{s_y^2/n_y + s_x^2/n_x}}$$

```
# Small sample test with unequal variance
t_test_unequal <- t.test(y, x, var.equal = FALSE)
t_test_unequal
#>
#>  Welch Two Sample t-test
#>
#> data:  y and x
#> t = 0.5, df = 8, p-value = 0.6305
#> alternative hypothesis: true difference in means is not
↪  equal to 0
#> 95 percent confidence interval:
#>  -3.612008  5.612008
#> sample estimates:
#> mean of x mean of y
#>        14        13
```

4.4.2 For Variances

To compare the variances of two independent samples, we can use the *F-test*. The test statistic is defined as

$$F_{ndf,ddf} = \frac{s_1^2}{s_2^2}$$

where $s_1^2 > s_2^2$, $ndf = n_1 - 1$, and $ddf = n_2 - 1$ are the numerator and denominator degrees of freedom, respectively.

4.4.2.1 F-Test

The hypotheses for the F-test are

$$H_0 : \sigma_y^2 = \sigma_x^2 \quad \text{(equal variances)}$$
$$H_a : \sigma_y^2 \neq \sigma_x^2 \quad \text{(unequal variances)}$$

The test statistic is

$$F = \frac{s_y^2}{s_x^2}$$

where s_y^2 and s_x^2 are the sample variances of the two groups.

Decision Rule
Reject H_0 if:

- $F > F_{n_y-1,n_x-1,\alpha/2}$ (upper critical value), or
- $F < F_{n_y-1,n_x-1,1-\alpha/2}$ (lower critical value)

Here:

- $F_{n_y-1,n_x-1,\alpha/2}$ and $F_{n_y-1,n_x-1,1-\alpha/2}$ are the critical points of the *F-distribution*, with $n_y - 1$ and $n_x - 1$ degrees of freedom.

Assumptions
- The F-test requires that the data in both groups follow a *normal distribution*.
- The F-test is sensitive to deviations from normality (e.g., heavy-tailed distributions). If the normality assumption is violated, it may lead to an inflated type I error rate (false positives).

Limitations and Alternatives:

1. **Sensitivity to Non-normality**:
   - When data have long-tailed distributions (positive kurtosis), the F-test may produce misleading results.
   - To assess normality, see Normality Assessment.
2. **Nonparametric Alternatives**:
   - If the normality assumption is not met, use robust tests such as the modified Levene's test (Brown-Forsythe test), which compares group variances based on medians instead of means.

```
# Load iris dataset
data(iris)

# Subset data for two species
irisVe <- iris$Petal.Width[iris$Species == "versicolor"]
irisVi <- iris$Petal.Width[iris$Species == "virginica"]

# Perform F-test
f_test <- var.test(irisVe, irisVi)

# Display results
f_test
#>
#>  F test to compare two variances
#>
#> data:  irisVe and irisVi
#> F = 0.51842, num df = 49, denom df = 49, p-value =
#> 0.02335
#> alternative hypothesis: true ratio of variances is not
↪  equal to 1
#> 95 percent confidence interval:
#>  0.2941935 0.9135614
#> sample estimates:
#> ratio of variances
#>          0.5184243
```

4.4.2.2 Levene's Test

Levene's test is a robust method for testing the equality of variances across multiple groups. Unlike the F-test, it is less sensitive to departures from normality and is particularly useful for handling non-normal distributions and datasets with outliers. The test works by analyzing the deviations of individual observations from their group mean or median.

Test Procedure

1. Compute the absolute deviations of each observation from its group mean or median:

   - For group y:

   $$d_{y,i} = |y_i - \text{Central Value}_y|.$$

   - For group x:

   $$d_{x,j} = |x_j - \text{Central Value}_x|.$$

   - The "central value" can be either the *mean* (classic Levene's test) or the *median* [modified Levene's test (Brown-Forsythe test) variation, more robust for non-normal data].

2. Perform a one-way ANOVA on the absolute deviations to test for differences in group variances.

Hypotheses

- Null Hypothesis (H_0): All groups have equal variances.
- Alternative Hypothesis (H_a): At least one group has a variance different from the others.

Test Statistic

The Levene's test statistic is calculated as an ANOVA on the absolute deviations. Let:

- k: the number of groups
- n_i: the number of observations in group i
- n: total number of observations

The test statistic is

$$W = \frac{(n-k)\sum_{i=1}^{k} n_i(\bar{d}_i - \bar{d})^2}{(k-1)\sum_{i=1}^{k}\sum_{j=1}^{n_i}(d_{i,j} - \bar{d}_i)^2}$$

where:

- $d_{i,j}$: absolute deviations within group i
- $\bar{d}_i$: mean of the absolute deviations for group i
- $\bar{d}$: overall mean of the absolute deviations

Under the null hypothesis, $W \sim F_{k-1,n-k}$.

Decision Rule

- Compute the test statistic W.
- Reject H_0 at significance level α if:

$$W > F_{k-1,n-k,\alpha}$$

```
# Load required package
library(car)
#> Loading required package: carData

# Perform Levene's Test (absolute deviations from the mean)
levene_test_mean <- leveneTest(Petal.Width ~ Species, data =
↪  iris)

# Perform Levene's Test (absolute deviations from the median)
levene_test_median <-
    leveneTest(Petal.Width ~ Species, data = iris, center =
    ↪  median)

# Display results
levene_test_mean
#> Levene's Test for Homogeneity of Variance (center = median)
#>        Df F value    Pr(>F)
#> group   2  19.892 2.261e-08 ***
#>       147
#> ---
#> Signif. codes:
#> 0 '***' 0.001 '**' 0.01 '*' 0.05 '.' 0.1 ' ' 1
levene_test_median
#> Levene's Test for Homogeneity of Variance (center = median)
#>        Df F value    Pr(>F)
#> group   2  19.892 2.261e-08 ***
#>       147
#> ---
#> Signif. codes:
#> 0 '***' 0.001 '**' 0.01 '*' 0.05 '.' 0.1 ' ' 1
```

The output includes:

- **Df**: Degrees of freedom for the numerator and denominator.
- **F-value**: The computed value of the test statistic W.
- **p-value**: The probability of observing such a value under the null hypothesis.

- If the p-value is less than α, reject H_0 and conclude that the group variances are significantly different.
- Otherwise, fail to reject H_0 and conclude there is no evidence of a difference in variances.

Advantages of Levene's Test:

1. **Robustness**:
    - Handles non-normal data and outliers better than the F-test.
2. **Flexibility**:
    - By choosing the center value (mean or median), it can adapt to different data characteristics:
        - Use the mean for symmetric distributions.
        - Use the median for non-normal or skewed data.
3. **Versatility**:
    - Applicable to comparing variances across more than two groups, unlike the modified Levene's test (Brown-Forsythe test), which is limited to two groups.

4.4.2.3 Modified Levene's Test (Brown-Forsythe Test)

The *modified Levene's test* is a robust alternative to the F-test for comparing variances between two groups. Instead of using squared deviations (as in the F-test), this test considers the *absolute deviations* from the median, making it less sensitive to non-normal data and long-tailed distributions. It is, however, still appropriate for normally distributed data.

For each sample, compute the absolute deviations from the median:

$$d_{y,i} = |y_i - y_{.5}| \quad \text{and} \quad d_{x,i} = |x_i - x_{.5}|$$

Let:

- $\bar{d}_y$ and $\bar{d}_x$ be the means of the absolute deviations for groups y and x, respectively.

The test statistic is

$$t_L^* = \frac{\bar{d}_y - \bar{d}_x}{s\sqrt{\frac{1}{n_y} + \frac{1}{n_x}}}$$

where the pooled variance s^2 is

$$s^2 = \frac{\sum_{i=1}^{n_y}(d_{y,i} - \bar{d}_y)^2 + \sum_{j=1}^{n_x}(d_{x,j} - \bar{d}_x)^2}{n_y + n_x - 2}$$

Assumptions

1. **Constant Variance of Error Terms**:
   The test assumes equal error variances in each group under the null hypothesis.
2. **Moderate Sample Size**:
   The approximation $t_L^* \sim t_{n_y+n_x-2}$ holds well for moderate or large sample sizes.
   Decision Rule:

- Compute t_L^* using the formula above.
- Reject the null hypothesis of equal variances if:

$$|t_L^*| > t_{n_y+n_x-2;\alpha/2}$$

This is equivalent to applying a two-sample t-test to the absolute deviations.

```
# Absolute deviations from the median dVe
    <- abs(irisVe
    - median(irisVe))
dVi <-
abs(irisVi -
median(irisVi))

# Perform t-test on absolute deviations
levene_test <- t.test(dVe, dVi, var.equal = TRUE)

# Display results
levene_test
#>
#>  Two Sample t-test
#>
#> data:  dVe and dVi
#> t = -2.5584, df = 98, p-value = 0.01205
#> alternative hypothesis: true difference in means is not
↪  equal to 0
#> 95 percent confidence interval:
#>  -0.12784786 -0.01615214
#> sample estimates:
```

```
#> mean of x mean of y
#>     0.154     0.226
```

For small sample sizes, use the unequal variance t-test directly on the original data as a robust alternative:

```
# Small sample t-test with unequal variances
small_sample_test <-
t.test(irisVe, irisVi,
var.equal = FALSE)

# Display results
small_sample_test
#>
#>  Welch Two Sample t-test
#>
#> data:  irisVe and irisVi
#> t = -14.625, df = 89.043, p-value < 2.2e-16
#> alternative hypothesis: true difference in means is not
↪  equal to 0
#> 95 percent confidence interval:
#>  -0.7951002 -0.6048998
#> sample estimates:
#> mean of x mean of y
#>     1.326     2.026
```

4.4.2.4 Bartlett's Test

The *Bartlett's test* is a statistical procedure for testing the equality of variances across multiple groups. It assumes that the data in each group are normally distributed and are sensitive to deviations from normality. When the assumption of normality holds, Bartlett's test is more powerful than Levene's test.

Hypotheses for Bartlett's Test

- Null Hypothesis (H_0): All groups have equal variances.
- Alternative Hypothesis (H_a): At least one group has a variance different from the others.

  The test statistic for Bartlett's test is

$$B = \frac{(n-k)\log(S_p^2) - \sum_{i=1}^{k}(n_i - 1)\log(S_i^2)}{1 + \frac{1}{3(k-1)}\left(\sum_{i=1}^{k}\frac{1}{n_i - 1} - \frac{1}{n-k}\right)}$$

where:

- k: the number of groups
- n_i: the number of observations in group i
- $n = \sum_{i=1}^{k} n_i$: the total number of observations
- S_i^2: sample variance of group i
- S_p^2: pooled variance, given by:

$$S_p^2 = \frac{\sum_{i=1}^{k}(n_i - 1)S_i^2}{n - k}$$

Under the null hypothesis, the test statistic $B \sim \chi^2_{k-1}$.

Assumptions

1. **Normality**: The data in each group must follow a normal distribution.
2. **Independence**: Observations within and between groups must be independent.
3. **Equal Sample Sizes (Optional)**: Bartlett's test is more robust if sample sizes are approximately equal.
   Decision Rule:

- Compute the test statistic B.
- Compare B to the critical value of the chi-Square distribution at α and $k - 1$ degrees of freedom.
- Reject H_0 if:

$$B > \chi^2_{k-1,\alpha}$$

Alternatively, use the p-value:

- Reject H_0 if the p-value $\leq \alpha$.

```
# Perform Bartlett's Test
bartlett_test <- bartlett.test(Petal.Width ~ Species, data =
↪  iris)

# Display results
bartlett_test
#>
#>  Bartlett test of homogeneity of variances
#>
#> data:  Petal.Width by Species
#> Bartlett's K-squared = 39.213, df = 2, p-value =
#> 3.055e-09
```

The output includes:

- **Bartlett's K-squared**: The value of the test statistic B.
- **df**: Degrees of freedom $(k - 1)$, where k is the number of groups.
- **p-value**: The probability of observing such a value of B under H_0.
- If the p-value is less than α, reject H_0 and conclude that the variances are significantly different across groups.
- If the p-value is greater than α, fail to reject H_0 and conclude that there is no significant evidence of variance differences.

Limitations of Bartlett's Test:

1. **Sensitivity to Non-normality**:
   Bartlett's test is highly sensitive to departures from normality. Even slight deviations can lead to misleading results.
2. **Not Robust to Outliers**:
   Outliers can disproportionately affect the test result.
3. **Alternatives**:
   If the normality assumption is violated, use robust alternatives like:
   - Levene's test (absolute deviations)
   - Modified Levene's test (Brown-Forsythe test) (median-based absolute deviations)

Advantages of Bartlett's Test:

- **High Power**: Bartlett's test is more powerful than robust alternatives when the normality assumption holds.
- **Simple Implementation**: The test is easy to perform and interpret.

4.4.3 Power

To evaluate the power of a test, we consider the situation where the variances are equal across groups:

$$\sigma_y^2 = \sigma_x^2 = \sigma^2$$

Under the assumption of equal variances, we take *equal sample sizes* from both groups, i.e., $n_y = n_x = n$.

Hypotheses for One-Sided Testing
We are testing

$$H_0 : \mu_y - \mu_x \leq 0 \quad \text{vs.} \quad H_a : \mu_y - \mu_x > 0$$

Test Statistic
The α level *z-test* rejects H_0 if the test statistic:

$$z = \frac{\bar{y} - \bar{x}}{\sigma\sqrt{\frac{2}{n}}} > z_\alpha$$

where:

- $\bar{y}$ and $\bar{x}$ are the sample means.
- σ is the common standard deviation.
- z_α is the critical value from the standard normal distribution.

Power Function
The *power* of the test, denoted as $\pi(\mu_y - \mu_x)$, is the probability of correctly rejecting H_0 when $\mu_y - \mu_x$ is some specified value. Under the alternative hypothesis, the power function is

$$\pi(\mu_y - \mu_x) = \Phi\left(-z_\alpha + \frac{\mu_y - \mu_x}{\sigma}\sqrt{\frac{n}{2}}\right)$$

where:

- Φ is the cumulative distribution function (CDF) of the standard normal distribution.
- $\frac{\mu_y - \mu_x}{\sigma}\sqrt{\frac{n}{2}}$ represents the standardized effect size.

Determining the Required Sample Size
To achieve a desired power of $1 - \beta$ when the true difference is δ (the smallest difference of interest), we solve for the required sample size n. The power equation is

$$\Phi\left(-z_\alpha + \frac{\delta}{\sigma}\sqrt{\frac{n}{2}}\right) = 1 - \beta$$

Rearranging for n, the required sample size is

$$n = \frac{2\sigma^2}{\delta^2}\left(z_\alpha + z_\beta\right)^2$$

where:

- σ: The common standard deviation
- z_α: The critical value for the type I error rate α (one-sided test)
- z_β: The critical value for the type II error rate β (related to power $1 - \beta$)
- δ: The minimum detectable difference between the means

```
# Parameters
alpha <- 0.05   # Significance level
beta <- 0.2     # Type II error rate (1 - Power = 0.2)
sigma <- 1      # Common standard deviation
delta <- 0.5    # Minimum detectable difference

# Critical values
z_alpha <- qnorm(1 - alpha)
z_beta <- qnorm(1 - beta)

# Sample size calculation
n <- (2 * sigma ^ 2 * (z_alpha + z_beta) ^ 2) / delta ^ 2

# Output the required sample size (per group)
ceiling(n)
#> [1] 50
```

Sample Size for Two-Sided Tests
For a *two-sided test*, replace z_α with $z_{\alpha/2}$ to account for the two-tailed critical region:

$$n = 2\left(\frac{\sigma(z_{\alpha/2} + z_\beta)}{\delta}\right)^2$$

This ensures that the test has the required power $1 - \beta$ to detect a difference of size δ between the means at significance level α.

Adjustment for the Exact t-Test
When conducting an exact *two-sample t-test* for small sample sizes, the sample size calculation involves the *noncentral t-distribution*. An approximate correction can be applied using the critical values from the t-distribution instead of the z-distribution. Alternatively, use the p-value: The adjusted sample size is

$$n^* = 2\left(\frac{\sigma(t_{2n-2;\alpha/2} + t_{2n-2;\beta})}{\delta}\right)^2$$

where:

- $t_{2n-2;\alpha/2}$: the critical value for the t-distribution with $2n - 2$ degrees of freedom for significance level $\alpha/2$
- $t_{2n-2;\beta}$: the critical value for the t-distribution with $2n - 2$ degrees of freedom for power $1 - \beta$

This correction adjusts for the increased variability of the t-distribution, especially important for small sample sizes.

```r
# Parameters
alpha <- 0.05     # Significance level
power <- 0.8      # Desired power
sigma <- 1        # Common standard deviation
delta <- 0.5      # Minimum detectable difference

# Calculate sample size for two-sided test
sample_size <-
    power.t.test(
        delta = delta,
        sd = sigma,
        sig.level = alpha,
        power = power,
        type = "two.sample",
        alternative = "two.sided"
    )

# Display results
sample_size
#>
#>      Two-sample t test power calculation
#>
#>               n = 63.76576
#>           delta = 0.5
#>              sd = 1
#>       sig.level = 0.05
#>           power = 0.8
#>     alternative = two.sided
#>
#> NOTE: n is number in *each* group
```

Key Insights

1. **Z-Test vs. T-Test**:
   For large samples, the normal approximation (z-test) works well. For small samples, the t-test correction using the t-distribution is essential.
2. **Effect of Power and Significance Level**:
   - Increasing power ($1 - \beta$) or decreasing α requires larger sample sizes.
   - A smaller minimum detectable difference (δ) also requires a larger sample size.
3. **Two-Sided Tests**:
   Two-sided tests require larger sample sizes compared to one-sided tests due to the split critical region.

Formula Summary

Test type	Formula for sample size
One-sided test	$n = 2\left(\frac{\sigma(z_\alpha + z_\beta)}{\delta}\right)^2$
Two-sided test	$n = 2\left(\frac{\sigma(z_{\alpha/2} + z_\beta)}{\delta}\right)^2$
Approximate t-test	$n^* = 2\left(\frac{\sigma(t_{2n-2;\alpha/2} + t_{2n-2;\beta})}{\delta}\right)^2$

4.4.4 *Matched Pair Designs*

In *matched pair designs*, two treatments are compared by measuring responses for the same subjects under both treatments. This ensures that the effects of subject-to-subject variability are minimized, as each subject serves as their own control.

We have two treatments, and the data are structured as follows (Table 4.6): Here:

- y_i represents the observation under Treatment A.
- x_i represents the observation under Treatment B.
- $d_i = y_i - x_i$ is the difference for subject i.

Assumptions

:

1. Observations y_i and x_i are measured for the same subjects, inducing correlation.
2. The differences d_i are independent and identically distributed (iid) and follow a normal distribution:

$$d_i \sim N(\mu_D, \sigma_D^2)$$

Mean and Variance of the Difference

The mean difference μ_D and the variance σ_D^2 are given by

$$\mu_D = E(y_i - x_i) = \mu_y - \mu_x$$

$$\sigma_D^2 = \text{Var}(y_i - x_i) = \text{Var}(y_i) + \text{Var}(x_i) - 2 \cdot \text{Cov}(y_i, x_i)$$

Table 4.6 Paired sample structure with treatment values and differences

Subject	Treatment A	Treatment B	Difference
1	y_1	x_1	$d_1 = y_1 - x_1$
2	y_2	x_2	$d_2 = y_2 - x_2$
...	...	...	...
n	y_n	x_n	$d_n = y_n - x_n$

- If the *covariance* between y_i and x_i is *positive* (a typical case), the variance of the differences σ_D^2 is *reduced* compared to the independent sample case.
- This is the key benefit of *matched pair designs*: Reduced variability increases the precision of estimates.

Sample Statistics
For the differences $d_i = y_i - x_i$:

- The sample mean of the differences:

$$\bar{d} = \frac{1}{n}\sum_{i=1}^{n} d_i = \bar{y} - \bar{x}.$$

- The sample variance of the differences:

$$s_d^2 = \frac{1}{n-1}\sum_{i=1}^{n}(d_i - \bar{d})^2.$$

Once the data are converted into differences d_i, the problem reduces to *one-sample inference*. We can use tests and confidence intervals (CIs) for the mean of a single sample.

Hypothesis Test
We test the following hypotheses:

$$H_0 : \mu_D = 0 \quad \text{vs.} \quad H_a : \mu_D \neq 0$$

The test statistic is

$$t = \frac{\bar{d}}{s_d/\sqrt{n}} \sim t_{n-1}$$

where n is the number of subjects:

- Reject H_0 at significance level α if:

$$|t| > t_{n-1,\alpha/2}$$

Confidence Interval
A $100(1-\alpha)\%$ confidence interval for μ_D is

$$\bar{d} \pm t_{n-1,\alpha/2} \cdot \frac{s_d}{\sqrt{n}}$$

```r
# Sample data
treatment_a <- c(85, 90, 78, 92, 88)
treatment_b <- c(80, 86, 75, 89, 85)

# Compute differences
differences <- treatment_a - treatment_b

# Perform one-sample t-test on the differences
t_test <- t.test(differences, mu = 0, alternative =
↪  "two.sided")

# Display results
t_test
#>
#>  One Sample t-test
#>
#> data:  differences
#> t = 9, df = 4, p-value = 0.0008438
#> alternative hypothesis: true mean is not equal to 0
#> 95 percent confidence interval:
#>  2.489422 4.710578
#> sample estimates:
#> mean of x
#>       3.6
```

The output includes:

- **t-statistic**: The calculated test statistic for the matched pairs.
- **p-value**: The probability of observing such a difference under the null hypothesis.
- **Confidence Interval**: The range of plausible values for the mean difference μ_D.
- If the p-value is less than α, reject H_0 and conclude that there is a significant difference between the two treatments.
- If the confidence interval does not include 0, this supports the conclusion of a significant difference.

Key Insights:

1. **Reduced Variability**: Positive correlation between paired observations reduces the variance of the differences, increasing test power.
2. **Use of Differences**: The paired design converts the data into a single-sample problem for inference.
3. **Robustness**: The paired t-test assumes normality of the differences d_i. For larger n, the central limit theorem ensures robustness to non-normality.

Matched pair designs are a powerful way to control for subject-specific variability, leading to more precise comparisons between treatments.

4.4.5 Nonparametric Tests for Two Samples

For matched pair designs or independent samples where normality cannot be assumed, we use *nonparametric tests*. These tests do not assume any specific distribution of the data and are robust alternatives to parametric methods.

Stochastic Order and Location Shift

Suppose Y and X are random variables with cumulative distribution functions (CDFs) F_Y and F_X. Then Y is *stochastically larger* than X if, for all real numbers u:

$$P(Y > u) \geq P(X > u) \quad \text{(equivalently, } F_Y(u) \leq F_X(u)\text{)}.$$

If the two distributions differ only in their *location parameters*, say θ_y and θ_x, then we can frame the relationship as

$$Y > X \quad \text{if} \quad \theta_y > \theta_x.$$

We test the following hypotheses:

- **Two-Sided Hypothesis**:

$$H_0 : F_Y = F_X \quad \text{vs.} \quad H_a : F_Y \neq F_X.$$

- **Upper One-Sided Hypothesis**:

$$H_0 : F_Y = F_X \quad \text{vs.} \quad H_a : F_Y < F_X.$$

- **Lower One-Sided Hypothesis**:

$$H_0 : F_Y = F_X \quad \text{vs.} \quad H_a : F_Y > F_X.$$

We generally avoid the completely nondirectional alternative $H_a : F_Y \neq F_X$ because it allows arbitrary differences between the distributions, without requiring one distribution to be stochastically larger than the other.

Nonparametric Tests

When the focus is on whether the two distributions differ *only in location parameters*, two equivalent nonparametric tests are commonly used:

1. Wilcoxon signed rank test
2. Mann-Whitney U-test

Both tests are mathematically equivalent and test whether one sample is systematically larger than the other.

4.4.5.1 Wilcoxon Rank Sum Test

The *Wilcoxon rank test* is a nonparametric test used to compare two independent samples to assess whether their distributions differ in location. It is based on the ranks of the combined observations rather than their actual values.

Procedure

1. **Combine and Rank Observations**:
   Combine all $n = n_y + n_x$ observations (from both groups) into a single dataset and rank them in ascending order. If ties exist, assign the average rank to tied values.
2. **Calculate Rank Sums**:
   Compute the sum of ranks for each group:

   - w_y: Sum of the ranks for group y (sample 1).
   - w_x: Sum of the ranks for group x (sample 2).
     By definition:

$$w_y + w_x = \frac{n(n+1)}{2}.$$

3. **Test Statistic**:
   The test focuses on the rank sum w_y. Reject H_0 if w_y is *large* (indicating y systematically has larger values) or equivalently, if w_x is *small*.
4. **Null Distribution**:
   Under H_0 (no difference between groups), all possible arrangements of ranks among y and x are equally likely. The total number of possible rank arrangements is

$$\frac{(n_y + n_x)!}{n_y!\, n_x!}.$$

5. **Computational Considerations**:

   - For small samples, the exact null distribution of the rank sums can be calculated.
   - For large samples, an *approximate normal distribution* can be used.

Hypotheses:

- Null Hypothesis (H_0): The two samples come from identical distributions.
- Alternative Hypothesis (H_a): The two samples come from different distributions, or one distribution is systematically larger.
- **Two-Sided Test**:

$$H_a : F_Y \neq F_X.$$

- **One-Sided Test**:

$$H_a : F_Y > F_X \quad \text{or} \quad H_a : F_Y < F_X.$$

```
# Subset data for two species irisVe
    <-
    iris$Petal.Width[iris$Species
    == "versicolor"]
irisVi <-
iris$Petal.Width[iris$Species
== "virginica"]

# Perform Wilcoxon Rank Test (approximate version, large
↪  sample)
wilcox_result <- wilcox.test(
    irisVe,
    irisVi,
    alternative = "two.sided", # Two-sided test
    conf.level = 0.95,         # Confidence level
    exact = FALSE,             # Approximate test for large
     ↪  samples
    correct = TRUE             # Apply continuity correction
)

# Display results
wilcox_result
#>
#>  Wilcoxon rank sum test with continuity correction
#>
#> data:  irisVe and irisVi
#> W = 49, p-value < 2.2e-16
#> alternative hypothesis: true location shift is not equal to
↪  0
```

The output of `wilcox.test` includes:

- **W**: The test statistic, which is the smaller of the two rank sums.
- **p-value**: The probability of observing such a difference in rank sums under H_0.
- **Alternative Hypothesis**: Specifies whether the test was one-sided or two-sided.
- **Confidence Interval** (if applicable): Provides a range for the difference in medians.

Decision Rule:

1. Reject H_0 at significance level α if the p-value $\leq \alpha$.

2. For large samples, compare the test statistic to a critical value from the normal approximation.

Key Features:

1. **Robustness**:
   The test does not require assumptions of normality and is robust to outliers.
2. **Distribution-Free**:
   It evaluates whether two samples differ in location without assuming a specific distribution.
3. **Rank-Based**:
   It uses the ranks of the observations, which makes it scale invariant (resistant to data transformation).

Computational Considerations:

- For small sample sizes, the *exact distribution* of the rank sums is used.
- For large sample sizes, the *normal approximation* with continuity correction is applied for computational efficiency.

4.4.5.2 Mann-Whitney U-Test

The *Mann-Whitney U-Test* is a nonparametric test used to compare two independent samples. It evaluates whether one sample tends to produce larger observations than the other, based on pairwise comparisons. The test does not assume normality and is robust to outliers.

Procedure:

1. **Pairwise Comparisons**:
   Compare each observation y_i from sample Y with each observation x_j from sample X:
   - Let u_y be the number of pairs where $y_i > x_j$.
   - Let u_x be the number of pairs where $y_i < x_j$.

   By definition:

$$u_y + u_x = n_y n_x$$

   where n_y is the sample size for group Y, and n_x is the sample size for group X.
2. **Test Statistic**:
   Reject H_0 if u_y is *large* (or equivalently, if u_x is *small*).

The *Mann-Whitney U-test* and *Wilcoxon rank sum test* are related through the rank sums:

$$u_y = w_y - \frac{n_y(n_y+1)}{2}, \quad u_x = w_x - \frac{n_x(n_x+1)}{2}$$

Here, w_y and w_x are the rank sums for groups Y and X, respectively.
Hypotheses:

- Null Hypothesis (H_0): The two samples come from identical distributions.
- Alternative Hypothesis (H_a):
  - **Upper One-Sided**: $F_Y < F_X$ (sample Y is stochastically larger).
  - **Lower One-Sided**: $F_Y > F_X$ (sample X is stochastically larger).
  - **Two-Sided**: $F_Y \neq F_X$ (distributions differ in location).

Test Statistic for Large Samples

For large sample sizes n_y and n_x, the null distribution of U can be approximated by a normal distribution with:

- **Mean**:

$$E(U) = \frac{n_y n_x}{2}.$$

- **Variance**:

$$\text{Var}(U) = \frac{n_y n_x (n_y + n_x + 1)}{12}.$$

The standardized test statistic z is

$$z = \frac{u_y - \frac{n_y n_x}{2} - \frac{1}{2}}{\sqrt{\frac{n_y n_x (n_y + n_x + 1)}{12}}}.$$

The test rejects H_0 at level α if:

$$z \geq z_\alpha \quad \text{(one-sided)} \quad \text{or} \quad |z| \geq z_{\alpha/2} \quad \text{(two-sided)}.$$

For the *two-sided test*, we use

- $u_{\max} = \max(u_y, u_x)$.
- $u_{\min} = \min(u_y, u_x)$.

The p-value is given by

$$p\text{-value} = 2P(U \geq u_{\max}) = 2P(U \leq u_{\min}).$$

When $y_i = x_j$ (ties), assign a value of $1/2$ to both u_y and u_x for that pair. While the exact sampling distribution differs slightly when ties exist, the *large sample normal approximation* remains reasonable.

```
# Subset data for two species
irisVe <- iris$Petal.Width[iris$Species == "versicolor"]
irisVi <- iris$Petal.Width[iris$Species == "virginica"]

# Perform Mann-Whitney U Test
mann_whitney <- wilcox.test(
    irisVe, irisVi,
    alternative = "two.sided",
    conf.level = 0.95,
    exact = FALSE,    # Approximate test for large samples
    correct = TRUE    # Apply continuity correction
)

# Display results
mann_whitney
#>
#>  Wilcoxon rank sum test with continuity correction
#>
#> data:  irisVe and irisVi
#> W = 49, p-value < 2.2e-16
#> alternative hypothesis: true location shift is not equal to
↪  0
```

Decision Rule:

1. Reject H_0 if the p-value is less than α.
2. For large samples, check whether $z \ge z_{\alpha}$ (one-sided) or $|z| \ge z_{\alpha/2}$ (two-sided).

Key Insights:

1. **Robustness**: The Mann-Whitney U-test does not assume normality and is robust to outliers.
2. **Relationship to Wilcoxon Test**: The test is equivalent to the *Wilcoxon rank sum test* but formulated differently (based on pairwise comparisons).
3. **Large Sample Approximation**: For large n_y and n_x, the test statistic U follows an approximate normal distribution, simplifying computation.
4. **Handling Ties**: Ties are accounted for by assigning fractional contributions to u_y and u_x.

4.5 Categorical Data Analysis

Categorical data analysis is used when the outcome variables are *categorical*:

- **Nominal Variables**: Categories have no logical order (e.g., sex: male, female).
- **Ordinal Variables**: Categories have a logical order, but the relative distances between values are not well-defined (e.g., small, medium, large).

In categorical data, we often analyze how the distribution of one variable changes with the levels of another variable. For example, row percentages may differ across columns in a contingency table.

4.5.1 Association Tests

4.5.1.1 Small Samples

4.5.1.1.1 Fisher's Exact Test

For small samples, the approximate tests based on the asymptotic normality of $\hat{p}_1 - \hat{p}_2$ (the difference in proportions) do not hold. In such cases, we use *Fisher's exact test* to evaluate:

- Null Hypothesis (H_0): $p_1 = p_2$ (no association between variables).
- Alternative Hypothesis (H_a): $p_1 \neq p_2$ (an association exists).

Assumptions:

1. X_1 and X_2 are independent *binomial random variables*:
   - $X_1 \sim \text{Binomial}(n_1, p_1)$.
   - $X_2 \sim \text{Binomial}(n_2, p_2)$.
2. x_1 and x_2 are the observed values (successes in each sample).
3. Total sample size is $n = n_1 + n_2$.
4. Total successes are $m = x_1 + x_2$.

By conditioning on m, the total number of successes, the number of successes in sample 1 follows a *hypergeometric distribution*.

Test Statistic
To test $H_0 : p_1 = p_2$ against $H_a : p_1 \neq p_2$, we use the test statistic:

$$Z^2 = \left(\frac{\hat{p}_1 - \hat{p}_2}{\sqrt{\hat{p}(1-\hat{p})\left(\frac{1}{n_1} + \frac{1}{n_2}\right)}} \right)^2 \sim \chi^2_{1,\alpha}$$

where:

- $\hat{p}_1$ and $\hat{p}_2$ are the observed proportions of successes in samples 1 and 2.
- $\hat{p}$ is the pooled proportion:

$$\hat{p} = \frac{x_1 + x_2}{n_1 + n_2}.$$

- $\chi^2_{1,\alpha}$ is the upper α critical value of the *chi-squared distribution* with 1 degree of freedom.

Fisher's exact test can be extended to a *contingency table* setting to test whether the observed frequencies differ significantly from the expected frequencies under the null hypothesis of no association.

```
# Create a 2x2 contingency table
data_table <- matrix(c(8, 2, 1, 5), nrow = 2, byrow = TRUE)
colnames(data_table) <- c("Success", "Failure")
rownames(data_table) <- c("Group 1", "Group 2")

# Display the table
data_table
#>         Success Failure
#> Group 1       8       2
#> Group 2       1       5

# Perform Fisher's Exact Test
fisher_result <- fisher.test(data_table)

# Display the results
fisher_result
#>
#>  Fisher's Exact Test for Count Data
#>
#> data:  data_table
#> p-value = 0.03497
#> alternative hypothesis: true odds ratio is not equal to 1
#> 95 percent confidence interval:
#>     1.008849 1049.791446
#> sample estimates:
#> odds ratio
#>   15.46969
```

The output of `fisher.test()` includes:

- **p-value**: The probability of observing such a contingency table under the null hypothesis.

- **Alternative Hypothesis**: Indicates whether the test is two-sided or one-sided.
- If the p-value is less than α, reject H_0 and conclude that there is a significant association between the two variables.

4.5.1.1.2 Exact Chi-Square Test

For small samples where the normal approximation does not apply, we can compute the *exact chi-Square test* by using Fisher's exact test or Monte Carlo simulation methods.
The chi-square test statistic in the 2 × 2 table is
$\chi^2 = \sum_{i=1}^{r} \sum_{j=1}^{c} \frac{(O_{ij}-E_{ij})^2}{E_{ij}}$
where:

- O_{ij}: Observed frequency in cell (i, j)
- E_{ij}: Expected frequency under the null hypothesis
- r: Number of rows
- c: Number of columns

4.5.1.2 Large Samples

4.5.1.2.1 Pearson Chi-Square Test

The *Pearson chi-square test* is commonly used to test whether there is an association between two categorical variables. It compares the observed counts in a contingency table to the expected counts under the null hypothesis.
The test statistic is

$$\chi^2 = \sum_{\text{all cells}} \frac{(\text{observed} - \text{expected})^2}{\text{expected}}$$

The test is applied in settings where multiple proportions or frequencies are compared across independent surveys or experiments:

- **Null Hypothesis** (H_0): The observed data are consistent with the expected values (no association or no deviation from a model).
- **Alternative Hypothesis** (H_a): The observed data differ significantly from the expected values.

Characteristics of the Test:

1. **Validation of Models**:
   In some cases, H_0 represents the model whose validity is being tested. The goal is not necessarily to reject the model but to check whether the data are consistent with it. Deviations may be due to random chance.

2. **Strength of Association**:
   The chi-square test detects whether an association exists but does *not* measure the strength of the association. For measuring strength, metrics like Cramér's V or the phi coefficient should be used.
3. **Effect of Sample Size**:
   - The chi-square statistic reflects sample size. If the sample size is doubled (e.g., duplicating observations), the χ^2 statistic will also double, even though the strength of the association remains unchanged.
   - This sensitivity can sometimes lead to detecting significant results that are not practically meaningful.
4. **Expected Cell Frequencies**:
   - The test is not appropriate if more than *20% of the cells* in a contingency table have expected frequencies less than 5.
   - For small sample sizes, Fisher's exact test or exact p-values should be used instead.

1. **Test for a Single Proportion**:
   We test whether the observed proportion of successes equals 0.5.

$$H_0 : p_J = 0.5$$
$$H_a : p_J < 0.5$$

```
# Observed data
july.x <- 480
july.n <- 1000
# Test for single proportion
prop.test(
  x = july.x,
  n = july.n,
  p = 0.5,
  alternative = "less",
  correct = FALSE
)
#>
#>  1-sample proportions test without continuity
#>  correction
#>
#> data:  july.x out of july.n, null probability 0.5
#> X-squared = 1.6, df = 1, p-value = 0.103
#> alternative hypothesis: true p is less than 0.5
#> 95 percent confidence interval:
```

```
#>  0.0000000 0.5060055
#> sample estimates:
#>    p
#> 0.48
```

2. **Test for Equality of Proportions Between Two Groups**: We test whether the proportions of successes in July and September are equal (Table 4.7).

$$H_0 : p_J = p_S$$
$$H_a : p_j \neq p_S$$

```
# Observed data for two groups
sept.x <- 704
sept.n <- 1600
# Test for equality of proportions
prop.test(
  x = c(july.x, sept.x),
  n = c(july.n, sept.n),
  correct = FALSE
)
#>
#>  2-sample test for equality of proportions without
#>  continuity correction
#>
#> data:  c(july.x, sept.x) out of c(july.n, sept.n)
#> X-squared = 3.9701, df = 1, p-value = 0.04632
#> alternative hypothesis: two.sided
#> 95 percent confidence interval:
#>  0.0006247187 0.0793752813
#> sample estimates:
#> prop 1 prop 2
#>   0.48   0.44
```

Table 4.7 Comparison of proportions for multiple groups

	Experiment 1	Experiment 2	...	Experiment k
Number of successes	x_1	x_2	...	x_k
Number of failures	$n_1 - x_1$	$n_2 - x_2$	...	$n_k - x_k$
Total	n_1	n_2	...	n_k

Table 4.8 Expected counts for a goodness-of-fit test with estimated proportion

Success	$n_1\hat{p}$	$n_2\hat{p}$	...	$n_k\hat{p}$
Failure	$n_1(1-\hat{p})$	$n_2(1-\hat{p})$	...	$n_k(1-\hat{p})$
	n_1	n_2		n_k

Table 4.9 General form contingency table with marginal totals

	1	2	...	j	...	c	Row total
1	n_{11}	n_{12}	...	n_{1j}	...	n_{1c}	$n_{1.}$
2	n_{21}	n_{22}	...	n_{2j}	...	n_{2c}	$n_{2.}$
...	...	...	...	...	...	...	...
r	n_{r1}	n_{r2}	...	n_{rj}	...	n_{rc}	$n_{r.}$
Column total	$n_{.1}$	$n_{.2}$	...	$n_{.j}$	...	$n_{.c}$	$n_{..}$

We test the null hypothesis:

$$H_0 : p_1 = p_2 = \cdots = p_k$$

against the alternative that at least one proportion differs.

Pooled Proportion
Assuming H_0 is true, we estimate the common value of the probability of success as

$$\hat{p} = \frac{x_1 + x_2 + \cdots + x_k}{n_1 + n_2 + \cdots + n_k}.$$

The *expected counts* under H_0 are (Table 4.8):

The test statistic is

$$\chi^2 = \sum_{\text{all cells}} \frac{(\text{observed} - \text{expected})^2}{\text{expected}}$$

with $k - 1$ degrees of freedom.

Two-Way Contingency Tables
When categorical data are cross-classified, we create a two-way table of observed counts (Table 4.9).

Sampling Designs:

1. **Design 1: Total Sample Size Fixed**:
   - A single random sample of size n is drawn from the population.
   - Units are cross-classified into r rows and c columns. Both row and column totals are random variables.

- The cell counts n_{ij} follow a *multinomial distribution* with probabilities p_{ij} such that:

$$\sum_{i=1}^{r}\sum_{j=1}^{c} p_{ij} = 1.$$

- Let $p_{ij} = P(X = i, Y = j)$ be the joint probability, where X is the row variable and Y is the column variable.
- **Null Hypothesis of Independence**:

$$H_0 : p_{ij} = p_{i.}p_{.j}, \quad \text{where } p_{i.} = P(X = i) \text{ and } p_{.j} = P(Y = j).$$

- **Alternative Hypothesis**:

$$H_a : p_{ij} \neq p_{i.}p_{.j}.$$

2. **Design 2: Row Totals Fixed**:

- Random samples of sizes $n_1, n_2, \ldots, n_r$ are drawn independently from r row populations.
- The row totals $n_{i.}$ are fixed, but column totals are random.
- Counts in each row follow independent *multinomial distributions*.
- The null hypothesis assumes that the conditional probabilities of the column variable Y are the same across all rows:

$$H_0 : p_{ij} = P(Y = j|X = i) = p_j \quad \text{for all } i \text{ and } j.$$

- Alternatively:

$$H_0 : (p_{i1}, p_{i2}, \ldots, p_{ic}) = (p_1, p_2, \ldots, p_c) \quad \text{for all } i.$$

- **Alternative Hypothesis** (Table 4.10):

$$H_a : (p_{i1}, p_{i2}, \ldots, p_{ic}) \text{ are not the same for all } i.$$

Why Both Designs?

1. **Real-World Sampling Constraints**:

- Sometimes, you have control over row totals (e.g., fixed group sizes in stratified sampling).
- Other times, you collect data without predefined group sizes, and totals emerge randomly.

Table 4.10 Comparison of fixed total sample size vs. fixed row total designs in categorical data analysis

Design	Total sample size fixed	Row totals fixed
Scenario	A single dataset or experiment where all observations are collected together as one sample	Observations are collected separately for each row, with fixed totals for each row population
Example	Survey with 100 respondents randomly selected, recording responses based on two categorical variables (e.g., age group and gender)	Stratified survey with specific numbers of individuals sampled from predefined groups (e.g., 30 males, 40 females, 30 nonbinary)
Why This Design?	Models situations where the total number of observations is fixed. Both row and column categories emerge randomly	Models scenarios where sampling occurs independently within predefined strata or groups
	Tests for *independence* between two categorical variables (row and column)	Tests for *homogeneity* of column proportions across rows, ignoring differences in total counts between rows
Practical use case	*Market research*: Do customer demographics (rows) and purchase behavior (columns) show a dependence?	*Public Health*: Are smoking rates (columns) consistent across age groups (rows)?
	Biology: Is there an association between species (rows) and habitat types (columns)?	*Education*: Do pass rates (columns) differ across schools (rows), controlling for the number of students in each school?

2. **Different Null Hypotheses**:
   - Design 1 tests whether two variables are *independent* (e.g., does one variable predict the other?).
   - Design 2 tests whether column proportions are *homogeneous* across groups (e.g., are the groups similar?).

```
# Sampling Design 1: Total Sample Size Fixed
# Parameters for the multinomial distribution
r <- 3  # Number of rows
c <- 4  # Number of columns
n <- 100  # Total sample size
p <- matrix(c(0.1, 0.2, 0.1, 0.1,
              0.05, 0.15, 0.05, 0.1,
              0.05, 0.05, 0.025, 0.075), nrow = r, byrow =
              ↪ TRUE)

# Generate a single random sample
```

```r
set.seed(123)  # For reproducibility
n_ij <- rmultinom(1, size = n, prob = as.vector(p))

# Reshape into a contingency table
contingency_table_fixed_total <-
    matrix(n_ij,
           nrow = r,
           ncol = c,
           byrow = TRUE)
rownames(contingency_table_fixed_total) <- paste0("Row", 1:r)
colnames(contingency_table_fixed_total) <- paste0("Col", 1:c)

# Hypothesis testing (Chi-squared test of independence)
chisq_test_fixed_total <-
↪   chisq.test(contingency_table_fixed_total)
#> Warning in chisq.test(contingency_table_fixed_total):
#> Chi-squared approximation may be incorrect

# Display results
print("Contingency Table (Total Sample Size Fixed):")
#> [1] "Contingency Table (Total Sample Size Fixed):"
print(contingency_table_fixed_total)
#>      Col1 Col2 Col3 Col4
#> Row1    8    6    4   24
#> Row2   18    1    9    7
#> Row3    2    7    5    9
print("Chi-squared Test Results:")
#> [1] "Chi-squared Test Results:"
print(chisq_test_fixed_total)
#> 
#>  Pearson's Chi-squared test
#> 
#> data:  contingency_table_fixed_total
#> X-squared = 28.271, df = 6, p-value = 8.355e-05
```

- All counts in the contingency table come from a single multinomial sample where both row and column totals are random.
- **Conclusion**: Reject Null. The data suggests significant dependence between row and column variables.

```
# Sampling Design 2: Row Totals Fixed #
    Parameters for the fixed row totals n_row
    <-
c(30,
40, 30)
# Row totals c <-
4  # Number of columns p_col
<-
c(0.25,
0.25, 0.25,
0.25)  # Common column
probabilities under H0

# Generate independent multinomial samples for each row
set.seed(123)  # For reproducibility
row_samples <- lapply(n_row, function(size)
    t(rmultinom(1, size, prob = p_col)))

# Combine into a contingency table
contingency_table_fixed_rows <- do.call(rbind, row_samples)
rownames(contingency_table_fixed_rows) <- paste0("Row",
↪  1:length(n_row))
colnames(contingency_table_fixed_rows) <- paste0("Col", 1:c)

# Hypothesis testing (Chi-squared test of homogeneity)
chisq_test_fixed_rows <-
↪  chisq.test(contingency_table_fixed_rows)

# Display results
print("Contingency Table (Row Totals Fixed):")
#> [1] "Contingency Table (Row Totals Fixed):"
print(contingency_table_fixed_rows)
#>      Col1 Col2 Col3 Col4
#> Row1    6   10    7    7
#> Row2   13   13    4   10
#> Row3    8   10    6    6
print("Chi-squared Test Results:")
#> [1] "Chi-squared Test Results:"
print(chisq_test_fixed_rows)
#>
#>  Pearson's Chi-squared test
#>
#> data:  contingency_table_fixed_rows
#> X-squared = 3.2069, df = 6, p-value = 0.7825
```

- Row totals are fixed, and column counts within each row follow independent multinomial distributions.
- **Conclusion**: Fail to reject the null. The data does not provide evidence to suggest differences in column probabilities across rows.

Why Are the Results Different?

1. **Data Generation Differences**:
   - In *Design 1*, the entire table is treated as a single multinomial sample. This introduces dependencies between counts in the table.
   - In *Design 2*, rows are generated independently, and only the column probabilities are tested for consistency across rows.
2. **Null Hypotheses**:
   - *Design 1* tests independence between row and column variables (more restrictive).
   - *Design 2* tests homogeneity of column probabilities across rows (less restrictive).

Interpretation:

- The results are *not directly comparable* because the null hypotheses are different:
  - *Design 1* focuses on whether rows and columns are independent across the entire table.
  - *Design 2* focuses on whether column distributions are consistent across rows.
- **Real-World Implication**:
  - If you are testing for independence (e.g., whether two variables are unrelated), use Design 1.
  - If you are testing for consistency across groups (e.g., whether proportions are the same across categories), use Design 2.

Takeaways:

- The tests use the same statistical machinery (chi-squared test), but their interpretations differ based on the experimental design and null hypothesis.
- For the same dataset, differences in assumptions can lead to different conclusions.

4.5.1.2.2 Chi-Square Test for Independence

The expected frequencies $\hat{e}_{ij}$ under the null hypothesis are

$$\hat{e}_{ij} = \frac{n_{i.}n_{.j}}{n_{..}}$$

where $n_{i.}$ and $n_{.j}$ are the row and column totals, respectively, and $n_{..}$ is the total sample size.

The test statistic is

$$\chi^2 = \sum_{i=1}^{r} \sum_{j=1}^{c} \frac{(n_{ij} - \hat{e}_{ij})^2}{\hat{e}_{ij}} \sim \chi^2_{(r-1)(c-1)}$$

We reject H_0 at significance level α if

$$\chi^2 > \chi^2_{(r-1)(c-1),\alpha}.$$

Notes on the Pearson chi-square test:

- **Purpose**: Test for association or independence between two categorical variables.
- **Sensitivity to Sample Size**: The χ^2 statistic is proportional to sample size. Doubling the sample size doubles χ^2 even if the strength of the association remains unchanged.
- **Assumption on Expected Frequencies**: The test is not valid when more than 20% of the expected cell counts are less than 5. In such cases, exact tests are preferred.

```
# Create a contingency table
data_table <- matrix(c(30, 10, 20, 40), nrow = 2, byrow = TRUE)
colnames(data_table) <- c("Category 1", "Category 2")
rownames(data_table) <- c("Group 1", "Group 2")

# Display the table
print(data_table)
#>           Category 1 Category 2
#> Group 1           30         10
#> Group 2           20         40

# Perform Chi-Square Test
chi_result <- chisq.test(data_table)

# Display results
chi_result
#>
#>  Pearson's Chi-squared test with Yates' continuity
#>  correction
#>
#> data:  data_table
#> X-squared = 15.042, df = 1, p-value = 0.0001052
```

The output includes:

- Chi-Square Statistic (χ^2): The test statistic measuring the deviation between observed and expected counts.
- **p-value**: The probability of observing such a deviation under H_0.
- **Degrees of Freedom**: $(r - 1)(c - 1)$ for an $r \times c$ table.
- **Expected Frequencies**: The table of expected counts under H_0.
- If the p-value is less than α, reject H_0 and conclude that there is a significant association between the row and column variables.

4.5.1.3 Key Takeaways

See Table 4.11.

1. Fisher's exact test is specialized for small samples and fixed margins (2x2 tables)
2. Exact chi-square test is a broader version of Fisher's for larger tables but avoids asymptotic approximations.
3. Pearson chi-square test is the general framework, and its applications include:
   - Goodness-of-fit testing.
   - Testing independence (same as the chi-square test for independence).
4. Chi-square test for independence is a specific application of the Pearson chi-square test.

In essence:

- Fisher's exact test and exact chi-square test are precise methods for small datasets.
- Pearson chi-square test and chi-square test for independence are interchangeable terms in many contexts, focusing on larger datasets.

4.5.2 Ordinal Association

Ordinal association refers to a relationship between two variables where the levels of one variable exhibit a consistent pattern of increase or decrease in response to the levels of the other variable. This type of association is particularly relevant when dealing with ordinal variables, which have naturally ordered categories, such as ratings ("poor," "fair," "good," "excellent") or income brackets ("low," "medium," "high").
For example:

- As customer satisfaction ratings increase from "poor" to "excellent," the likelihood of recommending a product may also increase (positive ordinal association).

Table 4.11 Comparison of tests for categorical association

Test	Purpose	Key features	Sample size suitability	Statistical assumptions
Fisher's exact test	Tests association between two categorical variables in a **2** × **2** *table*	Computes exact *p*-values. Does not rely on asymptotic assumptions. Handles small sample sizes	*Small sample sizes*	Observations are independent. Fixed marginal totals. No normality assumption
Exact chi-square test	Tests association in larger contingency tables using exact methods	Generalization of Fisher's exact test. Avoids asymptotic assumptions. Suitable for small to medium datasets	*Small to medium sample sizes*	Observations are independent. Marginal totals may not be fixed. No normality assumption
Pearson chi-square test	Tests discrepancies between observed and expected frequencies	Most common chi-square-based test. Includes independence and goodness-of-fit tests. Relies on asymptotic assumptions	*Large sample sizes*	Observations are independent. Expected cell frequencies ≥ 5. Test statistic follows a chi-square distribution asymptotically
Chi-square test for independence	Tests independence between two categorical variables in a contingency table	Application of Pearson chi-square test. Same assumptions as asymptotic chi-square tests. Often used for larger contingency tables	*Medium to large sample sizes*	Observations are independent. Expected cell frequencies ≥ 5. Random sampling

- Alternatively, as stress levels move from "low" to "high," job performance may tend to decrease (negative ordinal association).

Key Characteristics of Ordinal Association:

1. **Logical Ordering of Levels**: The levels of both variables must follow a logical sequence. For instance, "small," "medium," and "large" are logically ordered, whereas categories like "blue," "round," and "tall" lack inherent order and are unsuitable for ordinal association.

2. **Monotonic Trends**: The association is typically monotonic, meaning that as one variable moves in a specific direction, the other variable tends to move in a consistent direction (either increasing or decreasing).
3. **Tests for Ordinal Association**: Specialized statistical tests assess ordinal association, focusing on how the rankings of one variable relate to those of the other. These tests require the data to respect the ordinal structure of both variables.

Practical Considerations: When using these tests, keep in mind:

- **Ordinal Data Handling**: Ensure that the data respects the ordinal structure (e.g., categories are correctly ranked and coded).
- **Sample Size**: Larger sample sizes provide more reliable estimates and stronger test power.
- **Contextual Relevance**: Interpret results within the context of the data and the research question. For example, a significant Spearman's correlation does not imply causation but rather a consistent trend.

4.5.2.1 Mantel-Haenszel Chi-Square Test

The Mantel-Haenszel chi-square test is a statistical tool for evaluating ordinal associations, particularly when the data consists of multiple 2×2 contingency tables that examine the same association under varying conditions or strata. Unlike measures of association such as correlation coefficients, this test does not quantify the strength of the association but rather evaluates whether an association exists after controlling for stratification.

The Mantel-Haenszel test is applicable to $2 \times 2 \times K$ contingency tables, where K represents the number of strata. Each stratum is a 2×2 table corresponding to different conditions or subgroups.

For each stratum k, let the marginal totals of the table be:

- $n_{.1k}$: Total observations in column 1
- $n_{.2k}$: Total observations in column 2
- $n_{1.k}$: Total observations in row 1
- $n_{2.k}$: Total observations in row 2
- $n_{..k}$: Total observations in the entire table

The observed cell count in row 1 and column 1 is denoted n_{11k}. Given the marginal totals, the sampling distribution of n_{11k} follows a hypergeometric distribution.

Under the assumption of conditional independence:

The expected value of n_{11k} is

$$m_{11k} = E(n_{11k}) = \frac{n_{1.k} n_{.1k}}{n_{..k}}$$

The variance of n_{11k} is

$$var(n_{11k}) = \frac{n_{1.k}n_{2.k}n_{.1k}n_{.2k}}{n_{..k}^2(n_{..k} - 1)}$$

Mantel and Haenszel proposed the test statistic:

$$M^2 = \frac{\left(|\sum_k n_{11k} - \sum_k m_{11k}| - 0.5\right)^2}{\sum_k var(n_{11k})} \sim \chi_1^2$$

where:

- The 0.5 adjustment, known as a continuity correction, improves the approximation to the χ^2 distribution.
- The test statistic follows a χ^2 distribution with 1 degree of freedom under the null hypothesis of conditional independence.

This method can be extended to general $I \times J \times K$ contingency tables, where I and J represent the number of rows and columns, respectively, and K is the number of strata.

Null Hypothesis (H_0):
There is no association between the two variables of interest across all strata, after controlling for the confounder.

In mathematical terms:

$$H_0 : \text{Odds Ratio (OR)} = 1 \text{ or Risk Ratio (RR)} = 1$$

Alternative Hypothesis (H_a):
There is an association between the two variables of interest across all strata, after controlling for the confounder.
In mathematical terms:

$$H_a : \text{Odds Ratio (OR)} \neq 1 \text{ or Risk Ratio (RR)} \neq 1$$

Let us consider a scenario where a business wants to evaluate the relationship between customer satisfaction (satisfied vs. not satisfied) and the likelihood of repeat purchases (yes vs. no) across different regions (e.g., North, South, and West). The goal is to determine whether this relationship holds consistently across the regions.

```
# Create a 2 x 2 x 3 contingency table
CustomerData = array(
    c(40, 30, 200, 300, 35, 20, 180, 265, 50, 25, 250, 275),
    dim = c(2, 2, 3),
    dimnames = list(
        Satisfaction = c("Satisfied", "Not Satisfied"),
        RepeatPurchase = c("Yes", "No"),
        Region = c("North", "South", "West")
    )
)

# View marginal table (summarized across regions)
margin.table(CustomerData, c(1, 2))
#>                RepeatPurchase
#> Satisfaction    Yes  No
#>   Satisfied     125 630
#>   Not Satisfied  75 840
```

1. Calculate the overall odds ratio (ignoring strata):

```
library(samplesizeCMH)
marginal_table = margin.table(CustomerData, c(1, 2))
odds.ratio(marginal_table)
#> [1] 2.222222
```

2. Calculate the conditional odds ratios for each region:

```
apply(CustomerData, 3, odds.ratio)
#>    North    South     West
#> 2.000000 2.576389 2.200000
```

3. The Mantel-Haenszel test evaluates whether the relationship between customer satisfaction and repeat purchases remains consistent across regions:

```
mantelhaen.test(CustomerData, correct = TRUE)
#>
#>  Mantel-Haenszel chi-squared test with continuity
#>  correction
#>
#> data:  CustomerData
#> Mantel-Haenszel X-squared = 26.412, df = 1, p-value =
#> 2.758e-07
#> alternative hypothesis: true common odds ratio is not equal
↪  to 1
```

```
#> 95 percent confidence interval:
#>  1.637116 3.014452
#> sample estimates:
#> common odds ratio
#>          2.221488
```

Interpretation:

1. **Overall Odds Ratio**: This provides an estimate of the overall association between satisfaction and repeat purchases, ignoring regional differences.
2. **Conditional Odds Ratios**: These show whether the odds of repeat purchases given satisfaction are similar across regions.
3. **Mantel-Haenszel Test**: A significant test result (e.g., $p < 0.05$) suggests that the relationship between satisfaction and repeat purchases is consistent across regions. Conversely, a nonsignificant result implies that regional differences may affect the association. By applying the Mantel-Haenszel test, businesses can determine if a marketing or customer retention strategy should be uniformly applied or customized to account for regional variations.

   1. There is strong evidence to suggest that the two variables of interest are associated across the strata (North, South, and West), even after accounting for potential confounding effects of stratification.
   2. The common odds ratio of approximately 2.22 indicates a substantial association, meaning that the outcome is more likely in the exposed group compared to the unexposed group.
   3. The variability in the stratum-specific odds ratios suggests that the strength of the association may differ slightly by region, but the Mantel-Haenszel test assumes the association is consistent (homogeneous).

4.5.2.2 McNemar's Test

McNemar's test is a special case of the Mantel-Haenszel chi-square test, designed for paired nominal data. It is particularly useful for evaluating changes in categorical responses before and after a treatment or intervention, or for comparing paired responses in matched samples. Unlike the Mantel-Haenszel test, which handles stratified data, McNemar's test is tailored to situations with a single 2×2 table derived from paired observations.

McNemar's test assesses whether the proportions of discordant pairs (off-diagonal elements in a 2×2 table) are significantly different. Specifically, it tests the null hypothesis that the probabilities of transitioning from one category to another are equal:

- **Null Hypothesis** (H_0):

$$P(\text{Switch from A to B}) = P(\text{Switch from B to A})$$

This implies that the probabilities of transitioning from one category to the other are equal, or equivalently, the off-diagonal cell counts (n_{12} and n_{21}) are symmetric:

$$H_0 : n_{12} = n_{21}.$$

- **Alternative Hypothesis** (H_A):

$$P(\text{Switch from A to B}) \neq P(\text{Switch from B to A})$$

This suggests that the probabilities of transitioning between categories are not equal, or equivalently, the off-diagonal cell counts (n_{12} and n_{21}) are asymmetric:

$$H_A : n_{12} \neq n_{21}$$

For example, consider a business analyzing whether a new advertising campaign influences customer preference for two products (A and B). Each customer is surveyed before and after the campaign, resulting in the following 2×2 contingency table:

- **Before rows**: preference for Product A or B before the campaign
- **After columns**: preference for Product A or B after the campaign

Let the table structure be (Table 4.12):

- n_{12}: customers who switched from Product A to B
- n_{21}: customers who switched from Product B to A

The test focuses on n_{12} and n_{21}, as they represent the discordant pairs.

The McNemar's test statistic is

$$M^2 = \frac{(|n_{12} - n_{21}| - 0.5)^2}{n_{12} + n_{21}}$$

where:

- The 0.5 is a continuity correction applied when sample sizes are small.

Table 4.12 Customer switching behavior between Product A and Product B

	After A	After B
Before A	n_{11}	n_{12}
Before B	n_{21}	n_{22}

- Under the null hypothesis of no preference change, M^2 follows a χ^2 distribution with 1 degree of freedom.

Let us analyze a voting behavior study where participants were surveyed before and after a campaign. The table represents:

- Rows: voting preference before the campaign (Yes, No)
- Columns: voting preference after the campaign (Yes, No)

```
# Voting preference before and after a campaign
vote = matrix(c(682, 22, 86, 810), nrow = 2, byrow = TRUE,
              dimnames = list(
                "Before" = c("Yes", "No"),
                "After" = c("Yes", "No")
              ))

# Perform McNemar's Test with continuity correction
mcnemar_result <- mcnemar.test(vote, correct = TRUE)
mcnemar_result
#>
#>  McNemar's Chi-squared test with continuity correction
#>
#> data:  vote
#> McNemar's chi-squared = 36.75, df = 1, p-value =
#> 1.343e-09
```

The test provides:

- **Test statistic** (M^2): Quantifies the asymmetry in discordant pairs.
- **p-value**: Indicates whether there is a significant difference in the discordant proportions.

Interpretation:

1. **Test Statistic**: A large M^2 value suggests significant asymmetry in the discordant pairs.
2. **p-value**:
    - A low p-value (e.g., $p < 0.05$) rejects the null hypothesis, indicating that the proportion of participants switching preferences (e.g., from Yes to No) is significantly different from those switching in the opposite direction (e.g., from No to Yes).
    - A high p-value fails to reject the null hypothesis, suggesting no significant preference change.

McNemar's test is widely used in business and other fields:

1. **Marketing Campaigns**: evaluating whether a campaign shifts consumer preferences or purchase intentions
2. **Product Testing**: determining if a new feature or redesign changes customer ratings
3. **Healthcare Studies**: analyzing treatment effects in paired medical trials

4.5.2.3 McNemar-Bowker Test

The McNemar-Bowker test is an extension of McNemar's test, designed for analyzing paired nominal data with more than two categories. It evaluates the symmetry of the full contingency table by comparing the off-diagonal elements across all categories. This test is particularly useful for understanding whether changes between categories are uniformly distributed or whether significant asymmetries exist.

Let the data be structured in an $r \times r$ square contingency table, where r is the number of categories, and the off-diagonal elements represent transitions between categories.

The hypotheses for the McNemar-Bowker test are:

- **Null Hypothesis** (H_0):

$$P(\text{Switch from Category } i \text{ to Category } j) = P(\text{Switch from Category } j \text{ to Category } i) \quad \forall i \neq j$$

  This implies that the off-diagonal elements are symmetric, and there is no directional preference in category transitions.
- **Alternative Hypothesis** (H_A):

$$P(\text{Switch from Category } i \text{ to Category } j) \neq P(\text{Switch from Category } j \text{ to Category } i) \quad \text{for at least one pair } (i, j)$$

  This suggests that the off-diagonal elements are not symmetric, indicating a directional preference in transitions between at least one pair of categories.

The McNemar-Bowker test statistic is

$$B^2 = \sum_{i<j} \frac{(n_{ij} - n_{ji})^2}{n_{ij} + n_{ji}}$$

where:

- n_{ij}: observed count of transitions from category i to category j
- n_{ji}: observed count of transitions from category j to category i

Under the null hypothesis, the test statistic B^2 approximately follows a χ^2 distribution with $\frac{r(r-1)}{2}$ degrees of freedom (corresponding to the number of unique pairs of categories).

For example, a company surveys customers about their satisfaction before and after implementing a new policy. Satisfaction is rated on a scale of 1 to 3 (1 = Low, 2 = Medium, 3 = High). The paired responses are summarized in the following 3 × 3 contingency table:

```
# Satisfaction ratings before and after the intervention
satisfaction_table <- matrix(c(
    30, 10, 5,  # Before: Low
    8, 50, 12,  # Before: Medium
    6, 10, 40   # Before: High
), nrow = 3, byrow = TRUE,
dimnames = list(
    "Before" = c("Low", "Medium", "High"),
    "After" = c("Low", "Medium", "High")
))

# Function to perform McNemar-Bowker Test
mcnemar_bowker_test <- function(table) {
  if (!all(dim(table)[1] == dim(table)[2])) {
    stop("Input must be a square matrix.")
  }

  # Extract off-diagonal elements
  n <- nrow(table)
  stat <- 0
  df <- 0

  for (i in 1:(n - 1)) {
    for (j in (i + 1):n) {
      nij <- table[i, j]
      nji <- table[j, i]
      stat <- stat + (nij - nji)^2 / (nij + nji)
      df <- df + 1
    }
  }

  p_value <- pchisq(stat, df = df, lower.tail = FALSE)
  return(list(statistic = stat, df = df, p_value = p_value))
```

```
}

# Run the test
result <- mcnemar_bowker_test(satisfaction_table)

# Print results
cat("McNemar-Bowker Test Results:\n")
#> McNemar-Bowker Test Results:
cat("Test Statistic (B^2):", result$statistic, "\n")
#> Test Statistic (B^2): 0.4949495
cat("Degrees of Freedom:", result$df, "\n")
#> Degrees of Freedom: 3
cat("p-value:", result$p_value, "\n")
#> p-value: 0.9199996
```

The output includes:

- **Test Statistic** (B^2): a measure of the asymmetry in the off-diagonal elements
- **p-value**: The probability of observing the data under the null hypothesis of symmetry

Interpretation:

1. **Test Statistic**: A large B^2 value suggests substantial asymmetry in transitions between categories.
2. **p-value**:
    - If the *p-value is less than the significance level* (e.g., $p < 0.05$), we reject the null hypothesis, indicating significant asymmetry in the transitions between at least one pair of categories.
    - If the *p-value is greater than the significance level*, we fail to reject the null hypothesis, suggesting that the category transitions are symmetric.

The McNemar-Bowker test has broad applications in business and other fields:

1. **Customer Feedback Analysis**: evaluating changes in customer satisfaction levels before and after interventions
2. **Marketing Campaigns**: assessing shifts in brand preferences across multiple brands in response to an advertisement
3. **Product Testing**: understanding how user preferences among different product features change after a redesign

4.5.2.4 Stuart-Maxwell Test

The Stuart-Maxwell test is used for analyzing changes in paired categorical data with more than two categories. It is a generalization of McNemar's test, applied to square contingency tables where the off-diagonal elements represent transitions between categories. Unlike the McNemar-Bowker test, which tests for symmetry across all pairs, the Stuart-Maxwell test focuses on overall marginal homogeneity.

The test evaluates whether the marginal distributions of paired data are consistent across categories. This is particularly useful when investigating whether the distribution of responses has shifted between two conditions, such as before and after an intervention.

Hypotheses for the Stuart-Maxwell Test:

- **Null Hypothesis** (H_0):

  The marginal distributions of the paired data are homogeneous (no difference).

- **Alternative Hypothesis** (H_A):

  The marginal distributions of the paired data are not homogeneous (there is a difference).

The Stuart-Maxwell test statistic is calculated as:

$$M^2 = \mathbf{b}'\mathbf{V}^{-1}\mathbf{b}$$

where:

- **b**: vector of differences between the marginal totals of paired categories
- **V**: covariance matrix of **b** under the null hypothesis

The test statistic M^2 follows a χ^2 distribution with $(r - 1)$ degrees of freedom, where r is the number of categories.

A company surveys employees about their satisfaction levels (low, medium, high) before and after implementing a new workplace policy. The results are summarized in the following 3×3 contingency table.

```
# Employee satisfaction data before and after a policy change
satisfaction_table <- matrix(c(
    40, 10, 5,   # Before: Low
    8, 50, 12,   # Before: Medium
    6, 10, 40    # Before: High
), nrow = 3, byrow = TRUE,
dimnames = list(
    "Before" = c("Low", "Medium", "High"),
    "After" = c("Low", "Medium", "High")
))

# Function to perform the Stuart-Maxwell Test
```

```
stuart_maxwell_test <- function(table) {
  if (!all(dim(table)[1] == dim(table)[2])) {
    stop("Input must be a square matrix.")
  }

  # Marginal totals for each category
  row_totals <- rowSums(table)
  col_totals <- colSums(table)

  # Vector of differences between row and column marginal
  ↪  totals
  b <- row_totals - col_totals

  # Covariance matrix under the null hypothesis
  total <- sum(table)
  V <- diag(row_totals + col_totals) -
       (outer(row_totals, col_totals, "+") / total)

  # Calculate the test statistic
  M2 <- t(b) %*% solve(V) %*% b
  df <- nrow(table) - 1
  p_value <- pchisq(M2, df = df, lower.tail = FALSE)

  return(list(statistic = M2, df = df, p_value = p_value))
}

# Run the Stuart-Maxwell Test
result <- stuart_maxwell_test(satisfaction_table)

# Print the results
cat("Stuart-Maxwell Test Results:\n")
#> Stuart-Maxwell Test Results:
cat("Test Statistic (M^2):", result$statistic, "\n")
#> Test Statistic (M^2): 0.01802387
cat("Degrees of Freedom:", result$df, "\n")
#> Degrees of Freedom: 2
cat("p-value:", result$p_value, "\n")
#> p-value: 0.9910286
```

Interpretation:

1. **Test Statistic**: Measures the extent of marginal differences in the table.
2. **p-value**:

- A *low p-value* (e.g., $p < 0.05$) indicates significant differences between the marginal distributions, suggesting a change in the distribution of responses.
- A *high p-value* suggests no evidence of marginal differences, meaning the distribution is consistent across conditions.

Practical Applications of the Stuart-Maxwell Test:

1. **Employee Surveys**: analyzing shifts in satisfaction levels before and after policy changes
2. **Consumer Studies**: evaluating changes in product preferences before and after a marketing campaign
3. **Healthcare Research**: assessing changes in patient responses to treatments across categories

4.5.2.5 Cochran-Mantel-Haenszel (CMH) Test

The Cochran-Mantel-Haenszel (CMH) test is a generalization of the Mantel-Haenszel chi-square test. It evaluates the association between two variables, while controlling for the effect of a third stratifying variable. This test is particularly suited for ordinal data, allowing researchers to detect trends and associations across strata. The CMH test addresses scenarios where:

1. Two variables (e.g., exposure and outcome) are ordinal or nominal.
2. A third variable (e.g., a demographic or environmental factor) stratifies the data into K independent groups.

The test answers: Is there a consistent association between the two variables across the strata defined by the third variable?

The CMH test has three main variations depending on the nature of the data:

1. **Correlation Test for Ordinal Data**: Assesses whether there is a linear association between two ordinal variables across strata.
2. **General Association Test**: Tests for any association (not necessarily ordinal) between two variables while stratifying by a third.
3. **Homogeneity Test**: Checks whether the strength of the association between the two variables is consistent across strata.

Hypotheses:

- **Null Hypothesis** (H_0): There is no association between the two variables across all strata, or the strength of the association is consistent across strata.
- **Alternative Hypothesis** (H_A): There is an association between the two variables in at least one stratum, or the strength of the association varies across strata.

The CMH test statistic is:

$$CMH = \frac{\left(\sum_k (O_k - E_k)\right)^2}{\sum_k V_k}$$

where:

- O_k: observed counts in stratum k
- E_k: expected counts in stratum k, calculated under the null hypothesis
- V_k: variance of the observed counts in stratum k

The test statistic follows a χ^2 distribution with 1 degree of freedom under the null hypothesis.

A company evaluates whether sales performance (low, medium, high) is associated with product satisfaction (low, medium, high) across three experience levels (junior, mid-level, senior). The data is organized into a $3 \times 3 \times 3$ contingency table.

```
# Sales performance data
sales_data <- array(
    c(20, 15, 10, 12, 18, 15, 8, 12, 20,    # Junior
      25, 20, 15, 20, 25, 30, 10, 15, 20,   # Mid-level
      30, 25, 20, 28, 32, 35, 15, 20, 30), # Senior
    dim = c(3, 3, 3),
    dimnames = list(
        SalesPerformance = c("Low", "Medium", "High"),
        Satisfaction = c("Low", "Medium", "High"),
        ExperienceLevel = c("Junior", "Mid-level", "Senior")
    )
)

# Load the vcd package for the CMH test
library(vcd)
#> Loading required package: grid

# Perform CMH Test
cmh_result <- mantelhaen.test(sales_data, correct = FALSE)
cmh_result
#>
#>  Cochran-Mantel-Haenszel test
#>
#> data:  sales_data
#> Cochran-Mantel-Haenszel M^2 = 22.454, df = 4, p-value
#> = 0.0001627
```

Interpretation:

1. **Test Statistic**: A large CMH statistic suggests a significant association between sales performance and satisfaction after accounting for experience level.
2. **p-value**:
    - A *low p-value* (e.g., $p < 0.05$) indicates a significant association between the two variables across strata.

- A *high p-value* suggests no evidence of association or that the relationship is consistent across all strata.

Practical Applications of the CMH Test:

1. **Business Performance Analysis**: investigating the relationship between customer satisfaction and sales performance across different demographic groups
2. **Healthcare Studies**: assessing the effect of treatment (e.g., dosage) on outcomes, while controlling for patient characteristics (e.g., age groups)
3. **Educational Research**: analyzing the relationship between test scores and study hours, stratified by teaching method

4.5.2.6 Summary Table of Tests

The following table provides a concise guide on when and why to use each test (Table 4.13):

Table 4.13 Tests for categorical association

Test name	When to use	Key question addressed	Data requirements
Mantel-Haenszel chi-square test	When testing for association between two binary variables across multiple strata	Is there a consistent association across strata?	Binary variables in $2 \times 2 \times K$ tables.
McNemar's test	When analyzing marginal symmetry in paired binary data	Are the proportions of discordant pairs equal?	Paired binary responses (2×2 table)
McNemar-Bowker test	When testing for symmetry in paired nominal data with more than two categories	Are the off-diagonal elements symmetric across all categories?	Paired nominal data in $r \times r$ tables
Cochran-Mantel-Haenszel (CMH) test	When testing ordinal or general associations while controlling for a stratifying variable	Is there an association between two variables after stratification?	Ordinal or nominal data in $I \times J \times K$ tables
Stuart-Maxwell test	When analyzing marginal homogeneity in paired nominal data with more than two categories	Are the marginal distributions of paired data homogeneous?	Paired nominal data in $r \times r$ tables

How to Choose the Right Test

1. **Paired vs. Stratified Data**:
   - Use McNemar's test or McNemar-Bowker test for paired data.
   - Use Mantel-Haenszel chi-square test or Cochran-Mantel-Haenszel (CMH) test for stratified data.
2. **Binary vs. Multi-category Variables**:
   - Use McNemar's test for binary data.
   - Use McNemar-Bowker test or Stuart-Maxwell test for multi-category data.
3. **Ordinal Trends**:
   - Use the Cochran-Mantel-Haenszel (CMH) test if testing for ordinal associations, while controlling for a stratifying variable.

4.5.3 Ordinal Trend

When analyzing ordinal data, it is often important to determine whether a consistent trend exists between variables. Tests for trend are specifically designed to detect monotonic relationships where changes in one variable are systematically associated with changes in another. These tests are widely used in scenarios involving ordered categories, such as customer satisfaction ratings, income brackets, or educational levels.

The primary objectives of trend tests are:

1. **To detect monotonic relationships**: Determine if higher or lower categories of one variable are associated with higher or lower categories of another variable.
2. **To account for ordinal structure**: Leverage the inherent order in the data to provide more sensitive and interpretable results compared to tests designed for nominal data.

Key Considerations for Trend Tests:

1. **Data Structure**:
   - Ensure that the variables have a natural order and are treated as ordinal.
   - Verify that the trend test chosen matches the data structure (e.g., binary outcome vs. multilevel ordinal variables).
2. **Assumptions**:
   - Many tests assume monotonic trends, meaning that the relationship should not reverse direction.
3. **Interpretation**:
   - A significant result indicates the presence of a trend but does not imply causality.

- The direction and strength of the trend should be carefully interpreted in the context of the data.

4.5.3.1 Cochran-Armitage Test

The Cochran-Armitage test for trend is a statistical method designed to detect a linear trend in proportions across ordered categories of a predictor variable. It is particularly useful in $2 \times J$ contingency tables, where there is a binary outcome (e.g., success/failure) and an ordinal predictor variable with J ordered levels.

The Cochran-Armitage test evaluates whether the proportion of a binary outcome changes systematically across the levels of an ordinal predictor. This test leverages the ordinal nature of the predictor to enhance sensitivity and power compared to general chi-square tests.

Hypotheses

- **Null Hypothesis** (H_0):

  The proportion of the binary outcome is constant across the levels of the ordinal predictor.

- **Alternative Hypothesis** (H_A):

  There is a linear trend in the proportion of the binary outcome across the levels of the ordinal predictor

The Cochran-Armitage test statistic is calculated as

$$Z = \frac{\sum_{j=1}^{J} w_j (n_{1j} - N_j \hat{p})}{\sqrt{\hat{p}(1-\hat{p}) \sum_{j=1}^{J} w_j^2 N_j}}$$

where:

- n_{1j}: count of the binary outcome (e.g., “success”) in category j
- N_j: total number of observations in category j
- $\hat{p}$: Overall proportion of the binary outcome, calculated as:

$$\hat{p} = \frac{\sum_{j=1}^{J} n_{1j}}{\sum_{j=1}^{J} N_j}$$

- w_j: Score assigned to the jth category of the ordinal predictor, often set to j for equally spaced levels.

The test statistic Z follows a standard normal distribution under the null hypothesis.

Table 4.14 Marketing campaign success by income level in a contingency table

Income level	Success	Failure	Total
Low	20	30	50
Medium	35	15	50
High	45	5	50

Key Assumptions:

1. **Ordinal Predictor**: The categories of the predictor variable must have a natural order.
2. **Binary Outcome**: The response variable must be dichotomous (e.g., success/failure).
3. **Independent Observations**: Observations within and across categories are independent.

Let us consider a study examining whether the success rate of a marketing campaign varies across three income levels (low, medium, high). The data is structured in a 2×3 contingency table (Table 4.14):

```
# Data: Success and Failure counts by Income Level
income_levels <- c("Low", "Medium", "High")
success <- c(20, 35, 45)
failure <- c(30, 15, 5)
total <- success + failure

# Scores for ordinal levels (can be custom weights)
scores <- 1:length(income_levels)

# Cochran-Armitage Test
# Function to calculate Z statistic
cochran_armitage_test <- function(success, failure, scores) {
  N <- success + failure
  p_hat <- sum(success) / sum(N)
  weights <- scores

  # Calculate numerator
  numerator <- sum(weights * (success - N * p_hat))

  # Calculate denominator
  denominator <- sqrt(p_hat * (1 - p_hat) * sum(weights^2 * N))

  # Z statistic
  Z <- numerator / denominator
  p_value <- 2 * (1 - pnorm(abs(Z)))

  return(list(Z_statistic = Z, p_value = p_value))
```

```
}

# Perform the test
result <- cochran_armitage_test(success, failure, scores)

# Print results
cat("Cochran-Armitage Test for Trend Results:\n")
#> Cochran-Armitage Test for Trend Results:
cat("Z Statistic:", result$Z_statistic, "\n")
#> Z Statistic: 2.004459
cat("p-value:", result$p_value, "\n")
#> p-value: 0.04502088
```

Interpretation:

1. **Test Statistic** (Z):
   - The Z-value indicates the strength and direction of the trend.
   - Positive Z: Proportions increase with higher categories.
   - Negative Z: Proportions decrease with higher categories.
2. **p-value**:
   - A *low p-value* (e.g., $p < 0.05$) rejects the null hypothesis, indicating a significant linear trend.
   - A *high p-value* fails to reject the null hypothesis, suggesting no evidence of a trend.

Practical Applications:

1. **Marketing**: analyzing whether customer success rates vary systematically across income levels or demographics
2. **Healthcare**: evaluating the dose-response relationship between medication levels and recovery rates
3. **Education**: studying whether pass rates improve with higher levels of educational support

4.5.3.2 Jonckheere-Terpstra Test

The Jonckheere-Terpstra test is a nonparametric test designed to detect ordered differences between groups. It is particularly suited for ordinal data where both the predictor and response variables exhibit a monotonic trend. Unlike general nonparametric tests like the Kruskal-Wallis test, which assess any differences between groups, the Jonckheere-Terpstra test specifically evaluates whether the data follows a prespecified ordering.

The Jonckheere-Terpstra test determines whether:

1. There is a monotonic trend in the response variable across ordered groups of the predictor.
2. The data aligns with an a priori hypothesized order (e.g., group 1 < group 2 < group 3).

Hypotheses:

- **Null Hypothesis** (H_0):

  There is no trend in the response variable across the ordered groups.

- **Alternative Hypothesis** (H_A):

  The response variable exhibits a monotonic trend across the ordered groups.

The trend can be increasing, decreasing, or as otherwise hypothesized.

The Jonckheere-Terpstra test statistic is based on the number of pairwise comparisons (U) that are consistent with the hypothesized trend. For k groups:

1. Compare all possible pairs of observations across groups.
2. Count the number of pairs where the values are consistent with the hypothesized order.

The test statistic T is the sum of all pairwise comparisons:

$$T = \sum_{i<j} T_{ij}$$

where T_{ij} is the number of concordant pairs between groups i and j.

Under the null hypothesis, T follows a normal distribution with:

- Mean:

$$\mu_T = \frac{N(N-1)}{4}$$

- Variance:

$$\sigma_T^2 = \frac{N(N-1)(2N+1)}{24}$$

Where N is the total number of observations.

The standardized test statistic is:

$$Z = \frac{T - \mu_T}{\sigma_T}$$

Key Assumptions:

1. **Ordinal or Interval Data**: The response variable must be at least ordinal, and the groups must have a logical order.
2. **Independent Groups**: Observations within and between groups are independent.
3. **Consistent Hypothesis**: The trend (e.g., increasing or decreasing) must be specified in advance.

Let us consider a study analyzing whether customer satisfaction ratings (on a scale of 1 to 5) improve with increasing levels of service tiers (basic, standard, premium). The data is grouped by service tier, and we hypothesize that satisfaction ratings increase with higher service tiers.

```
# Example Data: Customer Satisfaction Ratings by Service Tier
satisfaction <- list(
  Basic = c(2, 3, 2, 4, 3),
  Standard = c(3, 4, 3, 5, 4),
  Premium = c(4, 5, 4, 5, 5)
)

# Prepare data
ratings <- unlist(satisfaction)
groups <- factor(rep(names(satisfaction), times =
↪  sapply(satisfaction, length)))

# Calculate pairwise comparisons
manual_jonckheere <- function(ratings, groups) {
  n_groups <- length(unique(groups))
  pairwise_comparisons <- 0
  total_pairs <- 0

  # Iterate over group pairs
  for (i in 1:(n_groups - 1)) {
    for (j in (i + 1):n_groups) {
      group_i <- ratings[groups == levels(groups)[i]]
      group_j <- ratings[groups == levels(groups)[j]]

      # Count concordant pairs
      for (x in group_i) {
        for (y in group_j) {
          if (x < y) pairwise_comparisons <-
          ↪  pairwise_comparisons + 1
          if (x == y) pairwise_comparisons <-
          ↪  pairwise_comparisons + 0.5
          total_pairs <- total_pairs + 1
        }
```

```
      }
    }
  }

  # Compute test statistic
  T <- pairwise_comparisons
  N <- length(ratings)
  mu_T <- total_pairs / 2
  sigma_T <- sqrt(total_pairs * (N + 1) / 12)

  Z <- (T - mu_T) / sigma_T
  p_value <- 2 * (1 - pnorm(abs(Z)))

  return(list(T_statistic = T, Z_statistic = Z, p_value =
   ↪ p_value))
}

# Perform the test
result <- manual_jonckheere(ratings, groups)

# Print results
cat("Jonckheere-Terpstra Test Results:\n")
#> Jonckheere-Terpstra Test Results:
cat("T Statistic (Sum of Concordant Pairs):",
 ↪ result$T_statistic, "\n")
#> T Statistic (Sum of Concordant Pairs): 49.5
cat("Z Statistic:", result$Z_statistic, "\n")
#> Z Statistic: 1.2
cat("p-value:", result$p_value, "\n")
#> p-value: 0.2301393
```

Interpretation:

1. **Test Statistic** (T):
   - Represents the sum of all pairwise comparisons consistent with the hypothesized order.
   - Includes 0.5 for tied pairs.
2. Z **Statistic**:
   - A standardized measure of the strength of the trend
   - Calculated using T, the expected value of T under the null hypothesis (μ_T), and the variance of T (σ_T^2)

3. **p-value**:
   - A *low p-value* (e.g., $p < 0.05$) rejects the null hypothesis, indicating a significant trend in the response variable across ordered groups.
   - A *high p-value* fails to reject the null hypothesis, suggesting no evidence of a trend.

Practical Applications:

1. **Customer Experience Analysis**: assessing whether customer satisfaction increases with higher service levels or product tiers.
2. **Healthcare Studies**: testing whether recovery rates improve with increasing doses of a treatment
3. **Education Research**: analyzing whether test scores improve with higher levels of educational intervention

4.5.3.3 Mantel Test for Trend

The Mantel test for trend is a statistical method designed to detect a linear association between two ordinal variables. It is an extension of the Mantel-Haenszel chi-square test and is particularly suited for analyzing trends in ordinal contingency tables, such as $I \times J$ tables where both variables are ordinal.

The Mantel test for trend evaluates whether an increasing or decreasing trend exists between two ordinal variables. It uses the ordering of categories to assess linear relationships, making it more sensitive to trends compared to general association tests like chi square.

Hypotheses:

- **Null Hypothesis** (H_0):

  There is no linear association between the two ordinal variables.

- **Alternative Hypothesis** (H_A):

  There is a significant linear association between the two ordinal variables.

The Mantel test is based on the Pearson correlation between the row and column scores in an ordinal contingency table. The test statistic is:

$$M = \frac{\sum_i \sum_j w_i w_j n_{ij}}{\sqrt{\sum_i w_i^2 n_{i\cdot} \sum_j w_j^2 n_{\cdot j}}}$$

where:

- n_{ij}: observed frequency in cell (i, j)
- $n_{i.}$: row marginal total for row i
- $n_{.j}$: column marginal total for column j
- w_i: score for the ith row
- Score for the jth column

The test statistic M is asymptotically normally distributed under the null hypothesis.

Key Assumptions:

1. **Ordinal Variables**: Both variables must have a natural order.
2. **Linear Trend**: Assumes a linear relationship between the scores assigned to the rows and columns.
3. **Independence**: Observations must be independent.

Let us consider a marketing study evaluating whether customer satisfaction levels (low, medium, high) are associated with increasing purchase frequency (low, medium, high).

```
# Customer satisfaction and purchase frequency data
data <- matrix(
  c(10, 5, 2, 15, 20, 8, 25, 30, 12),
  nrow = 3,
  byrow = TRUE,
  dimnames = list(
    Satisfaction = c("Low", "Medium", "High"),
    Frequency = c("Low", "Medium", "High")
  )
)

# Assign scores for rows and columns
row_scores <- 1:nrow(data)
col_scores <- 1:ncol(data)

# Compute Mantel statistic manually
mantel_test_manual <- function(data, row_scores, col_scores) {
  numerator <- sum(outer(row_scores, col_scores, "*") * data)
  row_marginals <- rowSums(data)
  col_marginals <- colSums(data)
  row_variance <- sum(row_scores^2 * row_marginals)
  col_variance <- sum(col_scores^2 * col_marginals)

  M <- numerator / sqrt(row_variance * col_variance)
  z_value <- M
```

```
  p_value <- 2 * (1 - pnorm(abs(z_value))) # Two-tailed test

  return(list(Mantel_statistic = M, p_value = p_value))
}

# Perform the Mantel Test
result <- mantel_test_manual(data, row_scores, col_scores)

# Display results
cat("Mantel Test for Trend Results:\n")
#> Mantel Test for Trend Results:
cat("Mantel Statistic (M):", result$Mantel_statistic, "\n")
#> Mantel Statistic (M): 0.8984663
cat("p-value:", result$p_value, "\n")
#> p-value: 0.368937
```

Interpretation:

1. **Test Statistic** (M):
   - Represents the strength and direction of the linear association.
   - Positive M: Increasing trend.
   - Negative M: Decreasing trend.
2. **p-value**:
   - A *low p-value* (e.g., $p < 0.05$) indicates a significant linear association.
   - A *high p-value* suggests no evidence of a trend.

Practical Applications:

1. **Marketing Analysis**: investigating whether satisfaction levels are associated with purchase behavior or loyalty
2. **Healthcare Research**: testing for a dose-response relationship between treatment levels and outcomes
3. **Social Sciences**: analyzing trends in survey responses across ordered categories

4.5.3.4 Chi-Square Test for Linear Trend

The chi-square test for linear trend is a statistical method used to detect a linear relationship between an ordinal predictor and a binary outcome. It is an extension of the chi-square test, designed specifically for ordered categories, making it more sensitive to linear trends in proportions compared to a general chi-square test of independence.

The chi-square test for linear trend evaluates whether the proportions of a binary outcome (e.g., success/failure) change systematically across ordered categories of a predictor variable. It is widely used in situations such as analyzing dose-response relationships or evaluating trends in survey responses.

Hypotheses

- **Null Hypothesis** (H_0): There is no linear trend in the proportions of the binary outcome across ordered categories.
- **Alternative Hypothesis** (H_A): There is a significant linear trend in the proportions of the binary outcome across ordered categories.

The test statistic is

$$X^2_{\text{trend}} = \frac{\left(\sum_{j=1}^{J} w_j (p_j - \bar{p}) N_j\right)^2}{\sum_{j=1}^{J} w_j^2 \bar{p}(1-\bar{p}) N_j}$$

where: J: number of ordered categories. w_j: scores assigned to the jth category (typically $j = 1, 2, \ldots, J$). p_j: proportion of success in the jth category. $\bar{p}$: overall proportion of success across all categories. N_j: total number of observations in the jth category.

The test statistic follows a chi-square distribution with one degree of freedom under the null hypothesis.

Key Assumptions

1. **Binary Outcome**: The response variable must be binary (e.g., success/failure).
2. **Ordinal Predictor**: The predictor variable must have a natural order.
3. **Independent Observations**: Data across categories must be independent.

Let us consider a study analyzing whether the proportion of customers who recommend a product increases with customer satisfaction levels (low, medium, high).

```
# Example Data: Customer Satisfaction and Recommendation
satisfaction_levels <- c("Low", "Medium", "High")
success <- c(20, 35, 50)  # Number of customers who recommend
↪  the product
failure <- c(30, 15, 10)  # Number of customers who do not
↪  recommend the product
total <- success + failure

# Assign ordinal scores
scores <- 1:length(satisfaction_levels)

# Calculate overall proportion of success
p_hat <- sum(success) / sum(total)
```

Interpretation:

1. **Chi-square statistic** (X^2_{trend}):
   - Indicates the strength of the linear trend in the proportions.
2. **p-value**:
   - A *low p-value* (e.g., $p < 0.05$) rejects the null hypothesis, indicating a significant linear trend.
   - A *high p-value* suggests no evidence of a linear trend.

Practical Applications:

1. **Marketing**: analyzing whether customer satisfaction levels predict product recommendations or repurchase intentions
2. **Healthcare**: evaluating dose-response relationships in clinical trials
3. **Education**: testing whether higher levels of intervention improve success rates

4.5.3.5 Key Takeways

Test	Purpose	Key Assumptions	Use Cases
Cochran-Armitage test	Tests for a linear trend in proportions across ordinal categories	Binary response variable. Predictor variable is ordinal	Evaluating dose-response relationships, comparing proportions across ordinal groups
Jonckheere-Terpstra test	Tests for a monotonic trend in a response variable across ordered groups	Response variable is continuous or ordinal. Predictor variable is ordinal	Comparing medians or distributions across ordinal groups, e.g., treatment levels
Mantel test for trend	Evaluates a linear association between an ordinal predictor and response	Ordinal variables. Linear trend expected	Determining trends in stratified or grouped data
Chi-square test for linear trend	Tests for linear trends in categorical data using contingency tables	Contingency table with ordinal predictor. Sufficient sample size (expected frequencies > 5)	Analyzing trends in frequency data, e.g., examining disease prevalence by age groups

4.6 Divergence Metrics and Tests for Comparing Distributions

Divergence metrics are powerful tools used to measure the similarity or dissimilarity between probability distributions. Unlike deviation and deviance statistics, divergence metrics focus on the broader relationships between entire distributions, rather than individual data points or specific model fit metrics. Let us clarify these differences:

- **Deviation Statistics**: Measure the difference between the realization of a variable and some reference value (e.g., the mean). Common statistics derived from deviations include:
  - Standard deviation
  - Average absolute deviation
  - Median absolute deviation
  - Maximum absolute deviation
- **Deviance Statistics**: Assess the goodness of fit of statistical models. These are analogous to the sum of squared residuals in ordinary least squares (OLS) but are generalized for use in cases with maximum likelihood estimation (MLE). Deviance statistics are frequently employed in generalized linear models (GLMs).

Divergence statistics differ fundamentally by focusing on statistical distances between entire probability distributions, rather than on individual data points or model errors.

1. Divergence Metrics
- **Definition**: Divergence metrics measure how much one probability distribution differs from another.
- **Key Properties**:
  - **Asymmetry**: Many divergence metrics, such as Kullback-Leibler (KL) divergence, are not symmetric (i.e., $D(P\|\|Q) \neq D(Q\|\|P)$).
  - **Nonmetric**: They do not necessarily satisfy the properties of a metric (e.g., symmetry, triangle inequality).
  - **Unitless**: Divergences are often expressed in terms of information (e.g., bits or nats).
- **When to Use**:
  - Use divergence metrics to assess the degree of mismatch between two probability distributions, especially in machine learning, statistical inference, or model evaluation.

Table 4.15 Comparison of divergence and distance metrics in distribution analysis

Aspect	Divergence metrics	Distance metrics
Symmetry	Often asymmetric (e.g., KL divergence)	Always symmetric (e.g., Wasserstein)
Triangle inequality	Not satisfied	Satisfied
Use case	Quantifying how different distributions are	Measuring the dissimilarity or "cost" of transformation

2. Distance Metrics

- **Definition**: Distance metrics measure the "distance" or dissimilarity between two objects, including probability distributions, datasets, or points in space.
- **Key Properties**:
  - **Symmetry**: $D(P, Q) = D(Q, P)$.
  - **Triangle Inequality**: $D(P, R) \leq D(P, Q) + D(Q, R)$.
  - **Non-negativity**: $D(P, Q) \geq 0$, with $D(P, Q) = 0$ only if $P = Q$.
- **When to Use**:
  - Use distance metrics to compare datasets, distributions, or clustering outcomes where symmetry and geometric properties are important (Table 4.15).

Applications of Divergence Metrics

Divergence metrics have found wide utility across domains, including:

- **Detecting Data Drift in Machine Learning**: Used to monitor whether the distribution of incoming data differs significantly from training data.
- **Feature Selection**: Employed to identify features with the most distinguishing power by comparing their distributions across different classes.
- **Variational Autoencoders (VAEs)**: Divergence metrics (such as Kullback-Leibler divergence) are central to the loss functions used in training VAEs.
- **Reinforcement Learning**: Measures the similarity between policy distributions to improve decision-making processes.
- **Assessing Consistency**: Compares the distributions of two variables representing constructs to test their relationship or agreement.

Divergence metrics are also highly relevant in business settings, providing insights and solutions for a variety of applications, such as:

- **Customer Segmentation and Targeting**: Compares the distributions of customer demographics or purchase behavior across market segments to identify key differences and target strategies more effectively.
- **Market Basket Analysis**: Measures divergence between distributions of product co-purchases across regions or customer groups to optimize product bundling and cross-selling strategies.

- **Marketing Campaign Effectiveness**: Evaluates whether the distribution of customer responses (e.g., click-through rates or conversions) differs significantly before and after a marketing campaign, providing insights into its success.
- **Fraud Detection**: Monitors divergence in transaction patterns over time to detect anomalies that may indicate fraudulent activities.
- **Supply Chain Optimization**: Compares demand distributions across time periods or regions to optimize inventory allocation and reduce stock-outs or overstocking.
- **Pricing Strategy Evaluation**: Analyzes the divergence between pricing and purchase distributions across products or customer segments to refine pricing models and improve profitability.
- **Churn Prediction**: Compares distributions of engagement metrics (e.g., frequency of transactions or usage time) between customers likely to churn and those who stay, to design retention strategies.
- **Financial Portfolio Analysis**: Assesses divergence between the expected returns and actual performance distributions of different asset classes to adjust investment strategies.

4.6.1 *Kolmogorov-Smirnov Test*

The Kolmogorov-Smirnov (KS) test is a *nonparametric test* used to determine whether two distributions differ significantly or whether a sample distribution matches a reference distribution. It is applicable to *continuous distributions* and is widely used in hypothesis testing and model evaluation.

Mathematical Definition
The KS statistic is defined as

$$D = \max |F_P(x) - F_Q(x)|$$

where:

- $F_P(x)$ is the cumulative distribution function (CDF) of the first distribution (or sample).
- $F_Q(x)$ is the CDF of the second distribution (or theoretical reference distribution).
- D measures the maximum vertical distance between the two CDFs.

Hypotheses:

- **Null Hypothesis** (H_0): The empirical distribution follows a specified distribution (or the two samples are drawn from the same distribution).
- **Alternative Hypothesis** (H_1): The empirical distribution does not follow the specified distribution (or the two samples are drawn from different distributions).

Properties of the KS Statistic:

- **Range**:

$$D \in [0, 1].$$

  - $D = 0$: Perfect match between the distributions.
  - $D = 1$: Maximum dissimilarity between the distributions.

- **Nonparametric Nature**: The KS test makes no assumptions about the underlying distribution of the data.

The KS test is useful in various scenarios, including:

1. Comparing two empirical distributions to evaluate similarity
2. Testing goodness of fit for a sample against a theoretical distribution
3. Detecting data drift or shifts in distributions over time
4. Validating simulation outputs by comparing them to real-world data

Example 1 Continuous Distributions

```
# Load necessary libraries
library(stats)

# Generate two sample distributions
set.seed(1)
sample_1 <- rnorm(100)          # Sample from a standard normal
↪  distribution
sample_2 <- rnorm(100, mean = 1)  # Sample with mean shifted
↪  to 1

# Perform Kolmogorov-Smirnov test
ks_test_result <- ks.test(sample_1, sample_2)
print(ks_test_result)
#>
#>  Asymptotic two-sample Kolmogorov-Smirnov test
#>
#> data:  sample_1 and sample_2
#> D = 0.36, p-value = 4.705e-06
#> alternative hypothesis: two-sided
```

This compares the CDFs of the two samples. The p-value indicates whether the null hypothesis (that the samples come from the same distribution) can be rejected.

Example 2 Discrete Data with Bootstrapped KS Test
For discrete data, a bootstrapped version of the KS test is often used to bypass the continuity requirement.

```
library(Matching)
#> Loading required package: MASS
#> ##
#> ##  Matching (Version 4.10-15, Build Date: 2024-10-14)
#> ##  See https://www.jsekhon.com for additional
↪  documentation.
#> ##  Please cite software as:
#> ##   Jasjeet S. Sekhon. 2011. ``Multivariate and Propensity
↪  Score Matching
#> ##   Software with Automated Balance Optimization: The
↪  Matching package for R.''
#> ##   Journal of Statistical Software, 42(7): 1-52.
#> ##

# Define two discrete samples
discrete_sample_1 <- c(0:10)
discrete_sample_2 <- c(0:10)

# Perform bootstrapped KS test
ks_boot_result <- ks.boot(Tr = discrete_sample_1, Co =
↪  discrete_sample_2)
print(ks_boot_result)
#> $ks.boot.pvalue
#> [1] 1
#>
#> $ks
#>
#>  Exact two-sample Kolmogorov-Smirnov test
#>
#> data:  Tr and Co
#> D = 0, p-value = 1
#> alternative hypothesis: two-sided
#>
#>
#> $nboots
#> [1] 1000
#>
#> attr(,"class")
#> [1] "ks.boot"
```

This method performs a bootstrapped version of the KS test, suitable for discrete data. The p-value indicates whether the null hypothesis (that the samples come from the same distribution) can be rejected.

Example 3 Comparing Multiple Distributions with KL Divergence (Optional Enhancement)
If you wish to extend the analysis to include divergence measures like KL divergence, use the following:

```
library(entropy)
library(tidyverse)
#> -- Attaching core tidyverse packages ---- tidyverse 2.0.0
↪  --
#> v dplyr     1.1.4     v readr     2.1.5
#> v forcats   1.0.0     v stringr   1.5.1
#> v ggplot2   3.5.2     v tibble    3.2.1
#> v lubridate 1.9.4     v tidyr     1.3.1
#> v purrr     1.0.4
#> -- Conflicts ---------------------- tidyverse_conflicts()
↪  --
#> x dplyr::filter() masks stats::filter()
#> x dplyr::lag()    masks stats::lag()
#> x dplyr::recode() masks car::recode()
#> x dplyr::select() masks MASS::select()
#> x purrr::some()   masks car::some()
#> i Use the conflicted package
↪  (<http://conflicted.r-lib.org/>) to force all conflicts to
↪  become errors

# Define multiple samples
lst <- list(sample_1 = c(1:20), sample_2 = c(2:30), sample_3 =
↪  c(3:30))

# Compute KL divergence between all pairs of distributions
result <- expand.grid(1:length(lst), 1:length(lst)) %>%
    rowwise() %>%
    mutate(KL = KL.empirical(lst[[Var1]], lst[[Var2]]))
#> Warning: There were 6 warnings in `mutate()`.
#> The first warning was:
#> i In argument: `KL = KL.empirical(lst[[Var1]],
#>   lst[[Var2]])`.
#> i In row 2.
#> Caused by warning in `freqs1 / freqs2`:
#> ! longer object length is not a multiple of shorter object
↪  length
#> i Run `dplyr::last_dplyr_warnings()` to see the 5 remaining
#>   warnings.
```

```
print(result)
#> # A tibble: 9 x 3
#> # Rowwise:
#>    Var1  Var2      KL
#>   <int> <int>   <dbl>
#> 1     1     1 0
#> 2     2     1 0.150
#> 3     3     1 0.183
#> 4     1     2 0.704
#> 5     2     2 0
#> 6     3     2 0.0679
#> 7     1     3 0.622
#> 8     2     3 0.0870
#> 9     3     3 0
```

This calculates the KL divergence for all pairs of distributions in the list, offering additional insights into the relationships between the distributions.

4.6.2 Anderson-Darling Test

The Anderson-Darling (AD) test is a *goodness-of-fit test* that evaluates whether a sample of data comes from a specific distribution. It is an enhancement of the Kolmogorov-Smirnov test, with greater sensitivity to deviations in the tails of the distribution. The Anderson-Darling test statistic is defined as

$$A^2 = -n - \frac{1}{n} \sum_{i=1}^{n} \left[(2i - 1) \left(\log F(Y_i) + \log(1 - F(Y_{n+1-i})) \right) \right]$$

where:

- n is the sample size.
- F is the cumulative distribution function (CDF) of the theoretical distribution being tested.
- Y_i are the ordered sample values.

The AD test modifies the basic framework of the KS test by giving more weight to the tails of the distribution, making it particularly sensitive to tail discrepancies.

Hypotheses

:

- **Null Hypothesis** (H_0): The sample data follows the specified distribution.

- **Alternative Hypothesis** (H_1): The sample data does not follow the specified distribution.

Key Properties:

1. **Tail Sensitivity**: Unlike the Kolmogorov-Smirnov test, the Anderson-Darling test emphasizes discrepancies in the tails of the distribution.
2. **Distribution-Specific Critical Values**: The AD test provides critical values tailored to the specific distribution being tested (e.g., normal, exponential).

The Anderson-Darling test is commonly used in:

1. Testing goodness of fit for a sample against theoretical distributions such as normal, exponential, or uniform
2. Evaluating the appropriateness of parametric models in hypothesis testing
3. Assessing distributional assumptions in quality control and reliability analysis

Example Testing Normality with the Anderson-Darling Test

```
library(nortest)

# Generate a sample from a normal distribution
set.seed(1)
sample_data <- rnorm(100, mean = 0, sd = 1)

# Perform the Anderson-Darling test for normality
ad_test_result <- ad.test(sample_data)
print(ad_test_result)
#>
#>  Anderson-Darling normality test
#>
#> data:  sample_data
#> A = 0.16021, p-value = 0.9471
```

If the p-value is below a chosen significance level (e.g., 0.05), the null hypothesis that the data is normally distributed is rejected.

Example Comparing Two Empirical Distributions
The AD test can also be applied to compare two empirical distributions using resampling techniques.

```
# Define two samples
set.seed(1)
sample_1 <- rnorm(100, mean = 0, sd = 1)
sample_2 <- rnorm(100, mean = 1, sd = 1)

# Perform resampling-based Anderson-Darling test
```

```
library(twosamples)
ad_test_result_empirical <- ad_test(sample_1, sample_2)
print(ad_test_result_empirical)
#>  Test Stat    P-Value
#> 6796.70454    0.00025
#> No bootstrap values were more extreme than the observed
↪  value.
#>  p-value = 1/(2*bootstraps) is an imprecise placeholder
```

This evaluates whether the two empirical distributions differ significantly.

4.6.3 Chi-Square Goodness-of-Fit Test

The *chi-square goodness-of-fit test* is a nonparametric statistical test used to evaluate whether a sample dataset comes from a population with a specific distribution. It compares observed frequencies with expected frequencies under a hypothesized distribution:

- **Null Hypothesis** (H_0): The data follow the specified distribution.
- **Alternative Hypothesis** (H_a): The data do not follow the specified distribution.

The chi-square test statistic is computed as

$$\chi^2 = \sum_{i=1}^{k} \frac{(O_i - E_i)^2}{E_i}$$

where:

- O_i: observed frequency for category i
- E_i: expected frequency for category i
- k: number of categories

The test statistic follows a chi-square distribution with degrees of freedom:

$$\nu = k - 1 - p$$

where p is the number of parameters estimated from the data.
Assumptions of the Test

1. **Random Sampling**: The sample data are drawn randomly from the population.
2. **Minimum Expected Frequency**: The expected frequencies E_i are sufficiently large (typically $E_i \geq 5$).
3. **Independence**: Observations in the sample are independent of each other.

Decision Rule:

1. Compute the test statistic χ^2 using the observed and expected frequencies.
2. Determine the critical value $\chi^2_{\alpha,\nu}$ for the chosen significance level α and degrees of freedom ν.
3. Compare χ^2 to $\chi^2_{\alpha,\nu}$:
   - Reject H_0 if $\chi^2 > \chi^2_{\alpha,\nu}$.
   - Alternatively, use the p-value approach:
     - Reject H_0 if $p \leq \alpha$.
     - Fail to reject H_0 if $p > \alpha$.

Steps for the chi-square goodness-of-fit test:

1. Define the expected frequencies based on the hypothesized distribution.
2. Compute the observed frequencies from the data.
3. Calculate the test statistic χ^2.
4. Determine the degrees of freedom ν.
5. Compare χ^2 with the critical value or use the p-value for decision-making.

Example (Testing a Fair Die) Suppose you are testing whether a six-sided die is fair. The die is rolled 60 times, and the observed frequencies of the outcomes are:

- **Observed Frequencies**: [10, 12, 8, 11, 9, 10].
- **Expected Frequencies**: A fair die has equal probability for each face, so $E_i = 60/6 = 10$ for each face.

```
# Observed frequencies
observed <- c(10, 12, 8, 11, 9, 10)

# Expected frequencies under a fair die
expected <- rep(10, 6)

# Perform Chi-Square Goodness-of-Fit Test
chisq_test <- chisq.test(x = observed, p = expected /
↪  sum(expected))

# Display results
chisq_test
#>
#>  Chi-squared test for given probabilities
#>
#> data:  observed
#> X-squared = 1, df = 5, p-value = 0.9626
```

Example Testing a Loaded Die For a die with unequal probabilities (e.g., a loaded die), the expected probabilities are defined explicitly:

```
# Observed frequencies
observed <- c(10, 12, 8, 11, 9, 10)

# Expected probabilities (e.g., for a loaded die)
probabilities <- c(0.1, 0.2, 0.3, 0.1, 0.2, 0.1)

# Expected frequencies
expected <- probabilities * sum(observed)

# Perform Chi-Square Goodness-of-Fit Test
chisq_test_loaded <- chisq.test(x = observed, p =
↪  probabilities)

# Display results
chisq_test_loaded
#>
#>  Chi-squared test for given probabilities
#>
#> data:  observed
#> X-squared = 15.806, df = 5, p-value = 0.007422
```

Limitations of the Chi-Square Test:

1. **Minimum Expected Frequency**: If $E_i < 5$ for any category, the test may lose power. Consider merging categories to meet this criterion.
2. **Independence**: Assumes observations are independent. Violations of this assumption can invalidate the test.
3. **Sample Size Sensitivity**: Large sample sizes may result in significant χ^2 values even for minor deviations from the expected distribution.

The chi-square goodness-of-fit test is a versatile tool for evaluating the fit of observed data to a hypothesized distribution, widely used in fields like quality control, genetics, and market research.

4.6.4 Cramér-von Mises Test

The *Cramér-von Mises (CvM) test* is a goodness-of-fit test that evaluates whether a sample dataset comes from a specified distribution. Similar to the Kolmogorov-Smirnov test (KS) and Anderson-Darling test (AD), it assesses the discrepancy between the empirical and theoretical cumulative distribution functions (CDFs).

However, the CvM test has *equal sensitivity across the entire distribution*, unlike the KS test (focused on the maximum difference) or the AD test (emphasizing the tails).

The Cramér-von Mises test statistic is defined as

$$W^2 = n \int_{-\infty}^{\infty} (F_n(x) - F(x))^2 \, dF(x)$$

where:

- n is the sample size.
- $F_n(x)$ is the empirical cumulative distribution function (ECDF) of the sample.
- $F(x)$ is the CDF of the specified theoretical distribution.

For practical implementation, the test statistic is often computed as

$$W^2 = \sum_{i=1}^{n} \left[F(X_i) - \frac{2i-1}{2n} \right]^2 + \frac{1}{12n}$$

where X_i are the ordered sample values.

Hypotheses
- **Null Hypothesis** (H_0): The sample data follow the specified distribution.
- **Alternative Hypothesis** (H_a): The sample data do not follow the specified distribution.

Key Properties
1. **Equal Sensitivity**:
   - The CvM test gives equal weight to discrepancies across all parts of the distribution, unlike the AD test, which emphasizes the tails.
2. **Nonparametric**:
   - The test makes no strong parametric assumptions about the data, aside from the specified distribution.
3. **Complementary to KS and AD tests**:
   - While the KS test focuses on the maximum distance between CDFs and the AD test emphasizes tails, the CvM test provides a balanced sensitivity across the entire range of the distribution.

The Cramér-von Mises test is widely used in:

1. **Goodness-of-Fit Testing**: assessing whether data follow a specified theoretical distribution (e.g., normal, exponential)

2. **Model Validation**: evaluating the fit of probabilistic models in statistical and machine learning contexts
3. **Complementary Testing**: Used alongside KS and AD tests for a comprehensive analysis of distributional assumptions

Example 1 Testing Normality

```
library(nortest)

# Generate a sample from a normal distribution
set.seed(1)
sample_data <- rnorm(100, mean = 0, sd = 1)

# Perform the Cramér-von Mises test for
    normality cvm_test_result <-
cvm.test(sample_data)
print(cvm_test_result)
#>  #>
Cramer-von Mises normality test #>
#> data:  sample_data
#> W = 0.026031, p-value = 0.8945
```

The test evaluates whether the sample data follow a normal distribution.

Example 2 Goodness of Fit for Custom Distributions
For distributions other than normal, you can use resampling techniques or custom implementations. Here is a pseudo-implementation for a custom theoretical distribution:

```
# Custom ECDF and theoretical CDF comparison
set.seed(1)
sample_data <-
    rexp(100, rate = 1)  # Sample from exponential
    ↪  distribution
theoretical_cdf <-
    function(x) {
        pexp(x, rate = 1)
    }  # Exponential CDF

# Compute empirical CDF
empirical_cdf <- ecdf(sample_data)

# Compute CvM statistic
cvm_statistic <-
```

```
    sum((empirical_cdf(sample_data) -
    ↪  theoretical_cdf(sample_data)) ^ 2) /
    length(sample_data)
print(paste("Cramér-von
Mises Statistic (Custom):",
            round(cvm_statistic, 4)))
#> [1] "Cramér-von Mises Statistic
(Custom): 0.0019"
```

This demonstrates a custom calculation of the CvM statistic for testing goodness of fit to an exponential distribution:

1. **Normality Test**:
   - The `cvm.test` function evaluates whether the sample data follow a normal distribution.
   - A small p-value indicates significant deviation from normality.
2. **Custom Goodness of Fit**:
   - Custom implementation allows testing for distributions other than normal.
   - The statistic measures the squared differences between the empirical and theoretical CDFs.

Advantages and Limitations:

1. **Advantages**:
   - Balanced sensitivity across the entire distribution.
   - Complements KS and AD tests by providing a different perspective on goodness of fit.
2. **Limitations**:
   - Critical values are distribution-specific.
   - The test may be less sensitive to tail deviations compared to the AD test.

The Cramér-von Mises test is a robust and versatile goodness-of-fit test, offering balanced sensitivity across the entire distribution. Its complementarity to KS and AD tests makes it an essential tool for validating distributional assumptions in both theoretical and applied contexts.

4.6.5 Kullback-Leibler Divergence

Kullback-Leibler (KL) divergence, also known as *relative entropy*, is a measure used to quantify the similarity between two probability distributions. It plays a critical role in statistical inference, machine learning, and information theory. However, KL divergence is not a true metric as it does not satisfy the triangle inequality.

Key Properties of KL Divergence

- **Not a Metric**: KL divergence fails to meet the triangle inequality requirement, and it is not symmetric, meaning:

$$D_{KL}(P\|Q) \neq D_{KL}(Q\|P).$$

- **Generalization to Multivariate Case**: KL divergence can be extended for multivariate distributions, making it flexible for complex analyses.
- **Quantifies Information Loss**: It measures the "information loss" when approximating the true distribution P with the predicted distribution Q. Thus, smaller values indicate closer similarity between the distributions.

Mathematical Definitions

KL divergence is defined differently for discrete and continuous distributions.

1. Discrete Case

For two discrete probability distributions $P = \{P_i\}$ and $Q = \{Q_i\}$, the KL divergence is given by

$$D_{KL}(P\|Q) = \sum_i P_i \log\left(\frac{P_i}{Q_i}\right)$$

2. Continuous Case

For continuous probability density functions $P(x)$ and $Q(x)$:

$$D_{KL}(P\|Q) = \int P(x) \log\left(\frac{P(x)}{Q(x)}\right) dx$$

- **Range**:

$$D_{KL}(P\|Q) \in [0, \infty).$$

  - $D_{KL} = 0$ indicates identical distributions ($P = Q$).
  - Larger values indicate greater dissimilarity between P and Q.

- **Nonsymmetric Nature**: As noted, $D_{KL}(P\|Q)$ and $D_{KL}(Q\|P)$ are not equal, emphasizing its directed nature.

```
library(philentropy)

# Example 1: Continuous case
# Define two continuous probability distributions with
↪  distinct patterns
X_continuous <- c(0.1, 0.2, 0.3, 0.4)  # Normalized to sum to
↪  1
Y_continuous <- c(0.4, 0.3, 0.2, 0.1)  # Normalized to sum to
↪  1

# Calculate KL divergence (logarithm base 2)
KL_continuous <- KL(rbind(X_continuous, Y_continuous), unit =
↪  "log2")
#> Metric: 'kullback-leibler' using unit: 'log2'; comparing: 2
↪  vectors.
print(paste("KL divergence (continuous):",
↪  round(KL_continuous, 2)))
#> [1] "KL divergence (continuous): 0.66"

# Example 2: Discrete case
# Define two discrete probability distributions
X_discrete <- c(5, 10, 15, 20)  # Counts for events
Y_discrete <- c(20, 15, 10, 5)  # Counts for events

# Estimate probabilities empirically and compute KL divergence
KL_discrete <- KL(rbind(X_discrete, Y_discrete), est.prob =
↪  "empirical")
#> Metric: 'kullback-leibler' using unit: 'log2'; comparing: 2
↪  vectors.
print(paste("KL divergence (discrete):", round(KL_discrete,
↪  2)))
#> [1] "KL divergence (discrete): 0.66"
```

Insights:

1. Continuous case uses normalized probability values explicitly provided.
2. Discrete case relies on empirical estimation of probabilities from counts.
3. Observe how KL divergence quantifies the “distance” between the two distributions.

4.6.6 Jensen-Shannon Divergence

Jensen-Shannon (JS) divergence is a symmetric and bounded measure of the similarity between two probability distributions. It is derived from the Kullback-Leibler divergence (KL), but addresses its asymmetry and unboundedness by incorporating a mixed distribution.

The Jensen-Shannon divergence is defined as:

$$D_{JS}(P\|Q) = \frac{1}{2}\left(D_{KL}(P\|M) + D_{KL}(Q\|M)\right)$$

where:

- $M = \frac{1}{2}(P + Q)$ is the *mixed distribution*, representing the average of P and Q.
- D_{KL} is the Kullback-Leibler divergence.

Key Properties:

- **Symmetry**: Unlike KL divergence, JS divergence is symmetric:

$$D_{JS}(P\|Q) = D_{JS}(Q\|P).$$

- **Boundedness**:
  - For base-2 logarithms:

$$D_{JS} \in [0, 1].$$

  - For natural logarithms (base-e):

$$D_{JS} \in [0, \ln(2)].$$

- **Interpretability**: The JS divergence measures the average information gain when moving from the mixed distribution M to either P or Q. Its bounded nature makes it easier to compare across datasets.

```
# Load the required library
library(philentropy)

# Example 1: Continuous case
# Define two continuous distributions
X_continuous <- 1:10  # Continuous sequence
Y_continuous <- 1:20  # Continuous sequence
```

```
# Compute JS divergence (logarithm base 2)
JS_continuous <- JSD(rbind(X_continuous, Y_continuous), unit =
↪ "log2")
#> Metric: 'jensen-shannon' using unit: 'log2'; comparing: 2
↪ vectors.
print(paste("JS divergence (continuous):",
↪ round(JS_continuous, 2)))
#> [1] "JS divergence (continuous): 20.03"

# X_continuous and Y_continuous represent continuous
↪ distributions.
# The mixed distribution (M) is computed internally as
# the average of the two distributions.

# Example 2: Discrete case
# Define two discrete distributions
X_discrete <- c(5, 10, 15, 20)  # Observed counts for events
Y_discrete <- c(20, 15, 10, 5)  # Observed counts for events

# Compute JS divergence with empirical probability estimation
JS_discrete <- JSD(rbind(X_discrete, Y_discrete), est.prob =
↪ "empirical")
#> Metric: 'jensen-shannon' using unit: 'log2'; comparing: 2
↪ vectors.
print(paste("JS divergence (discrete):", round(JS_discrete,
↪ 2)))
#> [1] "JS divergence (discrete): 0.15"

# X_discrete and Y_discrete represent event counts.
```

Probabilities are estimated empirically before calculating the divergence.

4.6.7 Hellinger Distance

The Hellinger distance is a bounded and symmetric measure of similarity between two probability distributions. It is widely used in statistics and machine learning to quantify how “close” two distributions are, with values ranging between 0 (identical distributions) and 1 (completely disjoint distributions).

Mathematical Definition
The Hellinger distance between two probability distributions P and Q is defined as

$$H(P, Q) = \frac{1}{\sqrt{2}}\sqrt{\sum_{x}\left(\sqrt{P(x)} - \sqrt{Q(x)}\right)^2}$$

where:

- $P(x)$ and $Q(x)$ are the probability densities or probabilities at point x for the distributions P and Q.
- The term $\sqrt{P(x)}$ is the square root of the probabilities, emphasizing geometric comparisons between the distributions.

Alternatively, for continuous distributions, the Hellinger distance can be expressed as

$$H(P, Q) = \frac{1}{\sqrt{2}}\sqrt{\int \left(\sqrt{P(x)} - \sqrt{Q(x)}\right)^2 dx}$$

Key Properties:

1. **Symmetry**:

$$H(P, Q) = H(Q, P).$$

   The distance is symmetric, unlike Kullback-Leibler divergence.
2. **Boundedness**:

$$H(P, Q) \in [0, 1].$$

   - $H = 0$: The distributions are identical ($P(x) = Q(x)$ for all x).
   - $H = 1$: The distributions have no overlap ($P(x) \neq Q(x)$).

3. **Interpretability**:

   - Hellinger distance provides a scale-invariant measure, making it suitable for comparing distributions in various contexts.

Hellinger distance is widely used in:

- **Hypothesis Testing**: comparing empirical distributions to theoretical models
- **Machine Learning**: feature selection, classification, and clustering tasks
- **Bayesian Analysis**: quantifying differences between prior and posterior distributions
- **Economics and Ecology**: measuring dissimilarity in distributions like income, species abundance, or geographical data

```
library(philentropy)

# Example 1: Compute Hellinger Distance for Discrete
↪  Distributions
# Define two discrete distributions as probabilities
P_discrete <- c(0.1, 0.2, 0.3, 0.4)  # Normalized
↪  probabilities
Q_discrete <- c(0.3, 0.3, 0.2, 0.2)  # Normalized
↪  probabilities

# Compute Hellinger distance
hellinger_discrete <-
    distance(rbind(P_discrete, Q_discrete), method =
     ↪  "hellinger")
#> Metric: 'hellinger'; comparing: 2 vectors.
print(paste("Hellinger Distance (Discrete):",
↪  round(hellinger_discrete, 4)))
#> [1] "Hellinger Distance (Discrete): 0.465"

# Example 2: Compute Hellinger Distance for Empirical
↪  Distributions
# Define two empirical distributions (counts)
P_empirical <- c(10, 20, 30, 40)  # Counts for distribution P
Q_empirical <- c(30, 30, 20, 20)  # Counts for distribution Q

# Normalize counts to probabilities
P_normalized <- P_empirical / sum(P_empirical)
Q_normalized <- Q_empirical / sum(Q_empirical)

# Compute Hellinger distance
hellinger_empirical <-
    distance(rbind(P_normalized, Q_normalized), method =
     ↪  "hellinger")
#> Metric: 'hellinger'; comparing: 2 vectors.
print(paste("Hellinger Distance (Empirical):",
↪  round(hellinger_empirical, 4)))
#> [1] "Hellinger Distance (Empirical): 0.465"
```

4.6.8 Bhattacharyya Distance

The *Bhattacharyya distance* is a statistical measure used to quantify the similarity or overlap between two probability distributions. It is commonly used in pattern recognition, signal processing, and statistics to evaluate how closely related two distributions are. The Bhattacharyya distance is particularly effective for comparing both discrete and continuous distributions.

The Bhattacharyya distance between two probability distributions P and Q is defined as

$$D_B(P, Q) = -\ln\left(\sum_x \sqrt{P(x)Q(x)}\right)$$

For continuous distributions, the Bhattacharyya distance is expressed as

$$D_B(P, Q) = -\ln\left(\int \sqrt{P(x)Q(x)}dx\right)$$

where:

- $P(x)$ and $Q(x)$ are the probability densities or probabilities for the distributions P and Q.
- The term $\int \sqrt{P(x)Q(x)}dx$ is known as the *Bhattacharyya coefficient*.

Key Properties:

1. **Symmetry**:

$$D_B(P, Q) = D_B(Q, P).$$

2. **Range**:

$$D_B(P, Q) \in [0, \infty).$$

   - $D_B = 0$: The distributions are identical ($P = Q$).
   - Larger values indicate less overlap and greater dissimilarity between P and Q.

3. **Relation to Hellinger Distance**:

   - The Bhattacharyya coefficient is related to the Hellinger distance:

$$H(P, Q) = \sqrt{1 - \sum_x \sqrt{P(x)Q(x)}}.$$

The Bhattacharyya distance is widely used in:

1. **Classification**: measuring the similarity between feature distributions in machine learning
2. **Hypothesis Testing**: evaluating the closeness of observed data to a theoretical model
3. **Image Processing**: comparing pixel intensity distributions or color histograms
4. **Economics and Ecology**: assessing similarity in income distributions or species abundance

Example 1 Discrete Distributions

```
# Define two discrete probability distributions
P_discrete <- c(0.1, 0.2, 0.3, 0.4)  # Normalized
↪  probabilities
Q_discrete <- c(0.3, 0.3, 0.2, 0.2)  # Normalized
↪  probabilities

# Compute Bhattacharyya coefficient
bhattacharyya_coefficient <- sum(sqrt(P_discrete *
↪  Q_discrete))

# Compute Bhattacharyya distance
bhattacharyya_distance <- -log(bhattacharyya_coefficient)

# Display results
print(paste(
    "Bhattacharyya Coefficient:",
    round(bhattacharyya_coefficient, 4)
))
#> [1] "Bhattacharyya Coefficient: 0.9459"
print(paste(
    "Bhattacharyya Distance (Discrete):",
    round(bhattacharyya_distance, 4)
))
#> [1] "Bhattacharyya Distance (Discrete): 0.0556"
```

A smaller Bhattacharyya distance indicates greater similarity between the two distributions.

Example 2 Continuous Distributions (Approximation)
For continuous distributions, the Bhattacharyya distance can be approximated using numerical integration or discretization.

```
# Generate two continuous distributions
set.seed(1)
P_continuous <-
    rnorm(1000, mean = 0, sd = 1)  # Standard normal
    ↪  distribution
Q_continuous <-
    rnorm(1000, mean = 1, sd = 1)  # Normal distribution with
    ↪  mean 1

# Create histograms to approximate probabilities
hist_P <- hist(P_continuous, breaks = 50, plot = FALSE)
hist_Q <- hist(Q_continuous, breaks = 50, plot = FALSE)

# Normalize histograms to probabilities
prob_P <- hist_P$counts / sum(hist_P$counts)
prob_Q <- hist_Q$counts / sum(hist_Q$counts)

# Compute Bhattacharyya coefficient
bhattacharyya_coefficient_continuous <- sum(sqrt(prob_P *
↪  prob_Q))

# Compute Bhattacharyya distance
bhattacharyya_distance_continuous <-
    -log(bhattacharyya_coefficient_continuous)

# Display results
print(paste(
    "Bhattacharyya Coefficient (Continuous):",
    round(bhattacharyya_coefficient_continuous, 4)
))
#> [1] "Bhattacharyya Coefficient (Continuous): 0.9823"
print(paste(
    "Bhattacharyya Distance (Continuous Approximation):",
    round(bhattacharyya_distance_continuous, 4)
))
#> [1] "Bhattacharyya Distance (Continuous Approximation):
↪  0.0178"
```

Continuous distributions are discretized into histograms to compute the Bhattacharyya coefficient and distance:

1. Discrete Case:

    1. The Bhattacharyya coefficient quantifies the overlap between P and Q.

   2. The Bhattacharyya distance translates this overlap into a logarithmic measure of dissimilarity.

2. Continuous Case:

   1. Distributions are discretized into histograms to approximate the Bhattacharyya coefficient and distance.

4.6.9 Wasserstein Distance

The Wasserstein distance, also known as the *Earth Mover's Distance (EMD)*, is a measure of similarity between two probability distributions. It quantifies the "cost" of transforming one distribution into another, making it particularly suitable for continuous data and applications where the geometry of the data matters.

Mathematical Definition
The Wasserstein distance between two probability distributions P and Q over a domain $\mathcal{X}$ is defined as

$$W_p(P, Q) = \left(\int_{\mathcal{X}} |F_P(x) - F_Q(x)|^p dx \right)^{\frac{1}{p}}$$

where:

- $F_P(x)$ and $F_Q(x)$ are the cumulative distribution functions (CDFs) of P and Q.
- $p \geq 1$ is the order of the Wasserstein distance (commonly $p = 1$).
- $|\cdot|^p$ is the absolute difference raised to the power p.

For the case of $p = 1$, the formula simplifies to

$$W_1(P, Q) = \int_{\mathcal{X}} |F_P(x) - F_Q(x)| dx$$

This represents the minimum "cost" of transforming the distribution P into Q, where cost is proportional to the distance a "unit of mass" must move.

Key Properties

- **Interpretability**: Represents the "effort" required to morph one distribution into another.
- **Metric**: Wasserstein distance satisfies the properties of a metric, including symmetry, non-negativity, and the triangle inequality.
- **Flexibility**: Can handle both empirical and continuous distributions.
  Wasserstein distance is widely used in various fields, including:

1. **Machine Learning**:
   - Training generative models such as Wasserstein GANs
   - Monitoring data drift in online systems
2. **Statistics**:
   - Comparing empirical distributions derived from observed data
   - Robustness testing under distributional shifts
3. **Economics**:
   - Quantifying disparities in income or wealth distributions
4. **Image Processing**:
   - Measuring structural differences between image distributions

```
library(transport)
library(twosamples)

# Example 1: Compute Wasserstein Distance (1D case)
set.seed(1)
# Generate a sample from a standard normal distribution
dist_1 <- rnorm(100)
# Generate a sample with mean shifted to 1
dist_2 <- rnorm(100, mean = 1)

# Calculate the Wasserstein distance
wass_distance <- wasserstein1d(dist_1, dist_2)
print(paste("1D Wasserstein Distance:", round(wass_distance,
↪ 4)))
#> [1] "1D Wasserstein Distance: 0.8533"

# Example 2: Wasserstein Metric as a Statistic
set.seed(1)
wass_stat_value <- wass_stat(dist_1, dist_2)
print(paste("Wasserstein Statistic:", round(wass_stat_value,
↪ 4)))
#> [1] "Wasserstein Statistic: 0.8533"

# Example 3: Wasserstein Test (Permutation-based Two-sample
↪ Test)
set.seed(1)
wass_test_result <- wass_test(dist_1, dist_2)
print(wass_test_result)
#> Test Stat   P-Value
```

```
#> 0.8533046 0.0002500
#> No bootstrap values were more extreme than the observed
↪   value.
#>  p-value = 1/(2*bootstraps) is an imprecise placeholder
```

- Example 1 calculates the simple Wasserstein distance between two distributions.
- Example 2 computes the Wasserstein distance as a statistical metric.
- Example 3 performs a permutation-based two-sample test using the Wasserstein metric.

4.6.10 Energy Distance

The *energy distance* is a statistical metric used to quantify the similarity between two probability distributions. It is particularly effective for comparing multidimensional distributions.

The energy distance between two distributions P and Q is defined as

$$E(P, Q) = 2\mathbb{E}[||X - Y||] - \mathbb{E}[||X - X'||] - \mathbb{E}[||Y - Y'||]$$

where:

- X and X' are independent and identically distributed (i.i.d.) random variables from P.
- Y and Y' are i.i.d. random variables from Q.
- $|| \cdot ||$ denotes the Euclidean distance.

Alternatively, for empirical distributions, the energy distance can be approximated as

$$E(P, Q) = \frac{2}{mn}\sum_{i=1}^{m}\sum_{j=1}^{n}||X_i - Y_j|| - \frac{1}{m^2}\sum_{i=1}^{m}\sum_{j=1}^{m}||X_i - X_j|| - \frac{1}{n^2}\sum_{i=1}^{n}\sum_{j=1}^{n}||Y_i - Y_j||$$

where:

- m and n are the sample sizes from distributions P and Q, respectively.
- X_i and Y_j are samples from P and Q.

Key Properties:

1. **Metric**:

   - Energy distance satisfies the properties of a metric: symmetry, non-negativity, and the triangle inequality.

2. **Range**:

$$E(P, Q) \geq 0.$$

- $E(P, Q) = 0$: The distributions are identical.
- Larger values indicate greater dissimilarity.

3. **Effectiveness for Multidimensional Data**:

- Energy distance is designed to work well in higher-dimensional spaces, unlike some traditional metrics.

The energy distance is widely used in:

1. **Hypothesis Testing**: testing whether two distributions are the same
2. **Energy Test** for equality of distributions
3. **Clustering**: measuring dissimilarity between clusters in multidimensional data
4. **Feature Selection**: comparing distributions of features across different classes to evaluate their discriminative power

Example 1 Comparing Two Distributions

```
# Load the 'energy' package
library(energy)

# Generate two sample distributions
set.seed(1)
X <- matrix(rnorm(1000, mean = 0, sd = 1), ncol = 2)  #
↪  Distribution P
Y <- matrix(rnorm(1000, mean = 1, sd = 1), ncol = 2)  #
↪  Distribution Q

# Combine X and Y and create a group identifier
combined <- rbind(X, Y)
groups <- c(rep(1, nrow(X)), rep(2, nrow(Y)))

# Compute Energy Distance
energy_dist <- edist(combined, sizes = table(groups))

# Print the Energy Distance
print(paste("Energy Distance:", round(energy_dist, 4)))
#> [1] "Energy Distance: 201.9202"
```

This calculates the energy distance between two multidimensional distributions.

Example 2: Energy Test for Equality of Distributions:

```
# Perform the Energy Test
energy_test <-
    eqdist.etest(rbind(X, Y), sizes = c(nrow(X), nrow(Y)), R =
    ↪  999)
print(energy_test)
#>
#>  Multivariate 2-sample E-test of equal distributions
#>
#> data:  sample sizes 500 500, replicates 999
#> E-statistic = 201.92, p-value = 0.001
```

The energy test evaluates the null hypothesis that the two distributions are identical:

1. **Energy Distance**:
   - Provides a single metric to quantify the dissimilarity between two distributions, considering all dimensions of the data.
2. **Energy Test**:
   - Tests for equality of distributions using energy distance.
   - The p-value indicates whether the distributions are significantly different.

Advantages of Energy Distance:

1. **Multidimensional Applicability**:
   - Works seamlessly with high-dimensional data, unlike some divergence metrics which may suffer from dimensionality issues.
2. **Nonparametric**:
   - Makes no assumptions about the form of the distributions.
3. **Robustness**:
   - Effective even with complex data structures.

4.6.11 Total Variation Distance

The *total variation (TV) distance* is a measure of the maximum difference between two probability distributions. It is widely used in probability theory, statistics, and machine learning to quantify how dissimilar two distributions are.

The total variation distance between two probability distributions P and Q is defined as

$$D_{TV}(P, Q) = \frac{1}{2} \sum_{x} |P(x) - Q(x)|$$

where:

- $P(x)$ and $Q(x)$ are the probabilities assigned to the outcome x by the distributions P and Q.
- The factor $\frac{1}{2}$ ensures that the distance lies within the range [0, 1].

Alternatively, for continuous distributions, the TV distance can be expressed as

$$D_{TV}(P, Q) = \frac{1}{2} \int |P(x) - Q(x)| dx$$

Key Properties:

1. **Range**:

$$D_{TV}(P, Q) \in [0, 1].$$

   - $D_{TV} = 0$: The distributions are identical ($P = Q$).
   - $D_{TV} = 1$: The distributions are completely disjoint (no overlap).

2. **Symmetry**:

$$D_{TV}(P, Q) = D_{TV}(Q, P).$$

3. **Interpretability**:
   - $D_{TV}(P, Q)$ represents the maximum probability mass that needs to be shifted to transform P into Q.

The total variation distance is used in:

1. **Hypothesis Testing**: quantifying the difference between observed and expected distributions
2. **Machine Learning**: evaluating similarity between predicted and true distributions
3. **Information Theory**: comparing distributions in contexts like communication and cryptography

Example 1 Discrete Distributions

```
# Define two discrete probability distributions
P_discrete <- c(0.1, 0.2, 0.3, 0.4)  # Normalized
↪  probabilities
Q_discrete <- c(0.3, 0.3, 0.2, 0.2)  # Normalized
↪  probabilities

# Compute Total Variation Distance
tv_distance <- sum(abs(P_discrete - Q_discrete)) / 2
print(paste("Total Variation Distance (Discrete):",
↪  round(tv_distance, 4)))
#> [1] "Total Variation Distance (Discrete): 0.3"
```

This calculates the maximum difference between the two distributions, scaled to lie between 0 and 1.

Example 2 Continuous Distributions (Approximation)
For continuous distributions, the TV distance can be approximated using discretization or numerical integration. Here is an example using random samples:

```
# Generate two continuous distributions
set.seed(1)
P_continuous <-
    rnorm(1000, mean = 0, sd = 1)  # Standard normal
    ↪  distribution
Q_continuous <-
    rnorm(1000, mean = 1, sd = 1)  # Normal distribution with
    ↪  mean 1

# Create histograms to approximate probabilities
hist_P <- hist(P_continuous, breaks = 50, plot = FALSE)
hist_Q <- hist(Q_continuous, breaks = 50, plot = FALSE)

# Normalize histograms to probabilities
prob_P <- hist_P$counts / sum(hist_P$counts)
prob_Q <- hist_Q$counts / sum(hist_Q$counts)

# Compute Total Variation Distance
tv_distance_continuous <- sum(abs(prob_P - prob_Q)) / 2
print(paste(
    "Total Variation Distance (Continuous Approximation):",
    round(tv_distance_continuous, 4)
))
#> [1] "Total Variation Distance (Continuous Approximation):
↪  0.125"
```

The continuous distributions are discretized into histograms, and TV distance is computed based on the resulting probabilities:

1. **Discrete Case**:
   - The TV distance quantifies the maximum difference between P and Q in terms of probability mass.
   - In this example, it highlights how much P and Q diverge.
2. **Continuous Case**:
   - For continuous distributions, TV distance is approximated using discretized probabilities from histograms.
   - This approach provides an intuitive measure of similarity for large samples.

The total variation distance provides an intuitive and interpretable measure of the maximum difference between two distributions. Its symmetry and bounded nature make it a versatile tool for comparing both discrete and continuous distributions.

4.6.12 Summary

See Tables 4.16, 4.17, and 4.18.

Table 4.16 Tests for comparing distributions

Test name	Purpose	Type of data	Advantages	Limitations
Kolmogorov-Smirnov test	Tests if two distributions are the same or if a sample matches a reference distribution	Empirical distributions (continuous)	Nonparametric, detects global differences	Less sensitive to tail differences, limited to one-dimensional data
Anderson-Darling test	Tests goodness of fit with emphasis on the tails	Continuous data	Strong sensitivity to tail behavior	Requires specifying a reference distribution
Chi-square goodness-of-fit test	Tests if observed frequencies match expected frequencies	Categorical data	Simple, intuitive for discrete data	Requires large sample sizes and sufficiently large expected frequencies
Cramér-von Mises test	Evaluates goodness of fit using cumulative distribution functions	Empirical distributions (continuous)	Sensitive across the entire distribution	Limited to one-dimensional data; requires cumulative distribution functions

Table 4.17 Divergence metrics

Metric name	Purpose	Type of data	Advantages	Limitations
Kullback-Leibler divergence	Measures how one probability distribution diverges from another	Probability distributions (continuous/discrete)	Provides a clear measure of information loss	Asymmetric, sensitive to zero probabilities
Jensen-Shannon divergence	Symmetric measure of similarity between two probability distributions	Probability distributions (continuous/discrete)	Symmetric and bounded; intuitive for comparison	Less sensitive to tail differences
Hellinger distance	Measures geometric similarity between two probability distributions	Discrete or continuous probability distributions	Easy to interpret; bounded between 0 and 1	Computationally expensive for large datasets
Bhattacharyya distance	Quantifies overlap between two statistical distributions	Probability distributions (continuous/discrete)	Useful for classification and clustering tasks	Less interpretable in large-scale applications

Table 4.18 Distance metrics

Metric name	Purpose	Type of data	Advantages	Limitations
Wasserstein distance	Measures the "effort" or "cost" to transform one distribution into another	Continuous or empirical distributions	Provides geometric interpretation; versatile	Computationally expensive for large-scale data
Energy distance	Measures statistical dissimilarity between multivariate distributions	Multivariate empirical distributions	Nonparametric, works well for high-dimensional data	Requires pairwise calculations; sensitive to outliers
Total variation distance	Measures the maximum absolute difference between probabilities of two distributions	Probability distributions (discrete/continuous)	Intuitive and strict divergence measure	Ignores structural differences beyond the largest deviation

Bibliography

Anderson TW, Darling DA (1952) Asymptotic theory of certain "goodness of fit" criteria based on stochastic processes. Ann Math Stat 23(2):193–212. http://www.jstor.org/stable/2236446

Bera AK, Jarque CM (1981) Efficient tests for normality, homoscedasticity and serial independence of regression residuals: Monte carlo evidence. Econ Lett 7(4):313–318

Geary RC (1936) Moments of the ratio of the mean deviation to the standard deviation for normal samples. Biometrika 28(3/4):295–307. https://doi.org/10.2307/2333953

Looney SW, Gulledge Jr TR (1985) Use of the correlation coefficient with normal probability plots. Am Stat 39(1):75–79

Marsaglia G, Marsaglia J (2004) Evaluating the anderson-darling distribution. J Stat Softw 9:1–5

Rubin DB (1996) Multiple imputation after 18+ years. J Am Stat Assoc 91(434):473–489

Satterthwaite FE (1946) An approximate distribution of estimates of variance components. Biom Bull 2(6):110–114

Shapiro SS, Francia R (1972) An approximate analysis of variance test for normality. J Am Stat Assoc 67(337):215–216

Shapiro SS, Wilk MB (1965) An analysis of variance test for normality (complete samples). Biometrika 52(3/4):591–611

Snedecor GW, Cochran WG (1989) Statistical methods (8th ed.). Iowa State University Press

Stephens MA (1974) Edf statistics for goodness of fit and some comparisons. J Am Stat Assoc 69(347):730–737

Student (1908) The probable error of a mean. Biometrika 6(1):1–25. https://doi.org/10.2307/2331554

M. Nguyen, *Foundations of Data Analysis*,
https://doi.org/10.1007/978-3-032-01858-8

MIX
Papier aus verantwortungsvollen Quellen
Paper from responsible sources
FSC® C105338

If you have any concerns about our products,
you can contact us on
ProductSafety@springernature.com

In case Publisher is established outside the EU,
the EU authorized representative is:
Springer Nature Customer Service Center GmbH
Europaplatz 3, 69115 Heidelberg, Germany

Printed by Libri Plureos GmbH
in Hamburg, Germany